Early Praise for *A Common-Sense Guide to Data Structures and Algorithms*

A Common-Sense Guide to Data Structures and Algorithms is a much-needed distillation of topics that elude many software professionals. The casual tone and presentation make it easy to understand concepts that are often hidden behind mathematical formulas and theory. This is a great book for developers looking to strengthen their programming skills.

➤ Jason Pike
Senior software engineer, Atlas RFID Solutions

At university, the "Data Structures and Algorithms" course was one of the driest in the curriculum; it was only later that I realized what a key topic it is. As a software developer, you must know this stuff. This book is a readable introduction to the topic that omits the obtuse mathematical notation common in many course texts.

➤ Nigel Lowry
Company director & principal consultant, Lemmata

Whether you are new to software development or a grizzled veteran, you will really enjoy and benefit from (re-)learning the foundations. Jay Wengrow presents a very readable and engaging tour through basic data structures and algorithms that will benefit every software developer.

➤ Kevin Beam
Software engineer, National Snow and Ice Data Center (NSIDC), University of Colorado Boulder

A Common-Sense Guide to Data Structures and Algorithms

Level Up Your Core Programming Skills

Jay Wengrow

The Pragmatic Bookshelf

Raleigh, North Carolina

Many of the designations used by manufacturers and sellers to distinguish their products are claimed as trademarks. Where those designations appear in this book, and The Pragmatic Programmers, LLC was aware of a trademark claim, the designations have been printed in initial capital letters or in all capitals. The Pragmatic Starter Kit, The Pragmatic Programmer, Pragmatic Programming, Pragmatic Bookshelf, PragProg and the linking *g* device are trademarks of The Pragmatic Programmers, LLC.

Every precaution was taken in the preparation of this book. However, the publisher assumes no responsibility for errors or omissions, or for damages that may result from the use of information (including program listings) contained herein.

Our Pragmatic books, screencasts, and audio books can help you and your team create better software and have more fun. Visit us at *https://pragprog.com*.

The team that produced this book includes:

Publisher: Andy Hunt
VP of Operations: Janet Furlow
Executive Editor: Susannah Davidson Pfalzer
Development Editor: Brian MacDonald
Copy Editor: Nicole Abramowtiz
Indexing: Potomac Indexing, LLC
Layout: Gilson Graphics

For sales, volume licensing, and support, please contact *support@pragprog.com*.

For international rights, please contact *rights@pragprog.com*.

Copyright © 2017 The Pragmatic Programmers, LLC.

All rights reserved. No part of this publication may be reproduced, stored in a retrieval system, or transmitted, in any form, or by any means, electronic, mechanical, photocopying, recording, or otherwise, without the prior consent of the publisher.

ISBN-13: 978-1-68050-244-2
Book version: P2.0—July 2018

Contents

	Preface	ix
1.	**Why Data Structures Matter**	1
	The Array: The Foundational Data Structure	2
	Reading	4
	Searching	7
	Insertion	9
	Deletion	11
	Sets: How a Single Rule Can Affect Efficiency	12
	Wrapping Up	15
2.	**Why Algorithms Matter**	17
	Ordered Arrays	18
	Searching an Ordered Array	20
	Binary Search	21
	Binary Search vs. Linear Search	24
	Wrapping Up	26
3.	**Oh Yes! Big O Notation**	27
	Big O: Count the Steps	28
	Constant Time vs. Linear Time	29
	Same Algorithm, Different Scenarios	31
	An Algorithm of the Third Kind	32
	Logarithms	33
	O(log N) Explained	34
	Practical Examples	35
	Wrapping Up	36
4.	**Speeding Up Your Code with Big O**	37
	Bubble Sort	37
	Bubble Sort in Action	38

Bubble Sort Implemented 42
The Efficiency of Bubble Sort 43
A Quadratic Problem 45
A Linear Solution 47
Wrapping Up 49

5. Optimizing Code with and Without Big O 51
Selection Sort 51
Selection Sort in Action 52
Selection Sort Implemented 56
The Efficiency of Selection Sort 57
Ignoring Constants 58
The Role of Big O 59
A Practical Example 61
Wrapping Up 62

6. Optimizing for Optimistic Scenarios 63
Insertion Sort 63
Insertion Sort in Action 64
Insertion Sort Implemented 68
The Efficiency of Insertion Sort 69
The Average Case 71
A Practical Example 74
Wrapping Up 76

7. Blazing Fast Lookup with Hash Tables 77
Enter the Hash Table 77
Hashing with Hash Functions 78
Building a Thesaurus for Fun and Profit, but Mainly Profit 79
Dealing with Collisions 82
The Great Balancing Act 85
Practical Examples 86
Wrapping Up 89

8. Crafting Elegant Code with Stacks and Queues . . . 91
Stacks 92
Stacks in Action 93
Queues 98
Queues in Action 100
Wrapping Up 101

9. Recursively Recurse with Recursion 103
Recurse Instead of Loop 103
The Base Case 105
Reading Recursive Code 105
Recursion in the Eyes of the Computer 108
Recursion in Action 110
Wrapping Up 112

10. Recursive Algorithms for Speed 113
Partitioning 113
Quicksort 118
The Efficiency of Quicksort 123
Worst-Case Scenario 126
Quickselect 128
Wrapping Up 131

11. Node-Based Data Structures 133
Linked Lists 133
Implementing a Linked List 135
Reading 136
Searching 137
Insertion 138
Deletion 140
Linked Lists in Action 142
Doubly Linked Lists 143
Wrapping Up 147

12. Speeding Up All the Things with Binary Trees 149
Binary Trees 149
Searching 152
Insertion 154
Deletion 157
Binary Trees in Action 163
Wrapping Up 165

13. Connecting Everything with Graphs 167
Graphs 168
Breadth-First Search 169
Graph Databases 178
Weighted Graphs 181

      Dijkstra's Algorithm   183
      Wrapping Up   189

14. Dealing with Space Constraints   **191**
      Big O Notation as Applied to Space Complexity   191
      Trade-Offs Between Time and Space   194
      Parting Thoughts   195

      Index   **197**

Preface

Data structures and algorithms are much more than abstract concepts. Mastering them enables you to write more efficient code that runs faster, which is particularly important for today's web and mobile apps. If you last saw an algorithm in a university course or at a job interview, you're missing out on the raw power algorithms can provide.

The problem with most resources on these subjects is that they're...well...obtuse. Most texts go heavy on the math jargon, and if you're not a mathematician, it's really difficult to grasp what on Earth is going on. Even books that claim to make algorithms "easy" assume that the reader has an advanced math degree. Because of this, too many people shy away from these concepts, feeling that they're simply not "smart" enough to understand them.

The truth, however, is that everything about data structures and algorithms boils down to common sense. Mathematical notation itself is simply a particular language, and everything in math can also be explained with common-sense terminology. In this book, I don't use any math beyond addition, subtraction, multiplication, division, and exponents. Instead, every concept is broken down in plain English, and I use a heavy dose of images to make everything a pleasure to understand.

Once you understand these concepts, you will be equipped to write code that is efficient, fast, and elegant. You will be able to weigh the pros and cons of various code alternatives, and be able to make educated decisions as to which code is best for the given situation.

Some of you may be reading this book because you're studying these topics at school, or you may be preparing for tech interviews. While this book will demystify these computer science fundamentals and go a long way in helping you at these goals, I encourage you to appreciate the power that these concepts provide in your day-to-day programming. I specifically go out of my way to make these concepts real and practical with ideas that you could make use of today.

Who Is This Book For?

This book is ideal for several audiences:

- You are a beginning developer who knows basic programming, but wants to learn the fundamentals of computer science to write better code and increase your programming knowledge and skills.

- You are a self-taught developer who has never studied formal computer science (or a developer who did but forgot everything!) and wants to leverage the power of data structures and algorithms to write more scalable and elegant code.

- You are a computer science student who wants a text that explains data structures and algorithms in plain English. This book can serve as an excellent supplement to whatever "classic" textbook you happen to be using.

- You are a developer who needs to brush up on these concepts since you may have not utilized them much in your career but expect to be quizzed on them in your next technical interview.

To keep the book somewhat language-agnostic, our examples draw from several programming languages, including Ruby, Python, and JavaScript, so having a basic understanding of these languages would be helpful. That being said, I've tried to write the examples in such a way that even if you're familiar with a different language, you should be able to follow along. To that end, I don't always follow the popular idioms for each language where I feel that an idiom may confuse someone new to that particular language.

What's in This Book?

As you may have guessed, this book talks quite a bit about data structures and algorithms. But more specifically, the book is laid out as follows:

In *Why Data Structures Matter* and *Why Algorithms Matter*, we explain what data structures and algorithms are, and explore the concept of time complexity—which is used to determine how efficient an algorithm is. In the process, we also talk a great deal about arrays, sets, and binary search.

In *Oh Yes! Big O Notation*, we unveil Big O Notation and explain it in terms that my grandmother could understand. We'll use this notation throughout the book, so this chapter is pretty important.

In *Speeding Up Your Code with Big O*, *Optimizing Code with and Without Big O*, and *Optimizing for Optimistic Scenarios*, we'll delve further into Big O Notation and use it practically to make our day-to-day code faster. Along the way, we'll cover various sorting algorithms, including Bubble Sort, Selection Sort, and Insertion Sort.

Blazing Fast Lookup with Hash Tables and *Crafting Elegant Code* discuss a few additional data structures, including hash tables, stacks, and queues. We'll show how these impact the speed and elegance of our code, and use them to solve real-world problems.

Recursively Recurse with Recursion introduces recursion, an anchor concept in the world of computer science. We'll break it down and see how it can be a great tool for certain situations. *Recursive Algorithms for Speed* will use recursion as the foundation for turbo-fast algorithms like Quicksort and Quickselect, and take our algorithm development skills up a few notches.

The following chapters, *Node-Based Data Structures*, *Speeding Up All the Things*, and *Connecting Everything with Graphs*, explore node-based data structures including the linked list, the binary tree, and the graph, and show how each is ideal for various applications.

The final chapter, *Dealing with Space Constraints*, explores space complexity, which is important when programming for devices with relatively small amounts of disk space, or when dealing with big data.

How to Read This Book

You've got to read this book in order. There are books out there where you can read each chapter independently and skip around a bit, but *this is not one of them*. Each chapter assumes that you've read the previous ones, and the book is carefully constructed so that you can ramp up your understanding as you proceed.

Another important note: to make this book easy to understand, I don't always reveal everything about a particular concept when I introduce it. Sometimes, the best way to break down a complex concept is to reveal a small piece of it, and only reveal the next piece when the first piece has sunken in. If I define a particular term as such-and-such, don't take that as the textbook definition until you've completed the entire section on that topic.

It's a trade-off: to make the book easy to understand, I've chosen to oversimplify certain concepts at first and clarify them over time, rather than ensure that every sentence is completely, academically, accurate. But don't worry too much, because by the end, you'll see the entire accurate picture.

Online Resources

This book has its own web page[1] on which you can find more information about the book, and help improve it by reporting errata, including content suggestions and typos.

You can find practice exercises for the content in each chapter at http://common-sensecomputerscience.com/, and in the code download package for this book.

Acknowledgments

While the task of writing a book may seem like a solitary one, this book simply could not have happened without the *many* people who have supported me in my journey writing it. I'd like to personally thank *all* of you.

To my wonderful wife, Rena—thank you for the time and emotional support you've given to me. You took care of everything while I hunkered down like a recluse and wrote. To my adorable kids—Tuvi, Leah, and Shaya—thank you for your patience as I wrote my book on "algorizms." And yes—it's finally finished.

To my parents, Mr. and Mrs. Howard and Debbie Wengrow—thank you for initially sparking my interest in computer programming and helping me pursue it. Little did you know that getting me a computer tutor for my ninth birthday would set the foundation for my career—and now this book.

When I first submitted my manuscript to the Pragmatic Bookshelf, I thought it was good. However, through the expertise, suggestions, and demands of all the wonderful people who work there, the book has become something much, much better than I could have written on my own. To my editor, Brian MacDonald—you've shown me how a book should be written, and your insights have sharpened each chapter; this book has your imprint all over it. To my managing editor, Susannah Pfalzer—you've given me the vision for what this book could be, taking my theory-based manuscript and transforming it into a book that can be applied to the everyday programmer. To the publishers Andy Hunt and Dave Thomas—thank you for believing in this book and making the Pragmatic Bookshelf the most wonderful publishing company to write for.

To the extremely talented software developer and artist Colleen McGuckin—thank you for taking my chicken scratch and transforming it into beautiful digital imagery. This book would be nothing without the spectacular visuals that you've created with such skill and attention to detail.

---

1. https://pragprog.com/book/jwdsal

I've been fortunate that so many experts have reviewed this book. Your feedback has been extremely helpful and has made sure that this book can be as accurate as possible. I'd like to thank all of you for your contributions: Aaron Kalair, Alberto Boschetti, Alessandro Bahgat, Arun S. Kumar, Brian Schau, Daivid Morgan, Derek Graham, Frank Ruiz, Ivo Balbaert, Jasdeep Narang, Jason Pike, Javier Collado, Jeff Holland, Jessica Janiuk, Joy McCaffrey, Kenneth Parekh, Matteo Vaccari, Mohamed Fouad, Neil Hainer, Nigel Lowry, Peter Hampton, Peter Wood, Rod Hilton, Sam Rose, Sean Lindsay, Stephan Kämper, Stephen Orr, Stephen Wolff, and Tibor Simic.

I'd also like to thank all the staff, students, and alumni at Actualize for your support. This book was originally an Actualize project, and you've all contributed in various ways. I'd like to particularly thank Luke Evans for giving me the idea to write this book.

Thank you all for making this book a reality.

Jay Wengrow
jay@actualize.co
August, 2017

CHAPTER 1

Why Data Structures Matter

Anyone who has written even a few lines of computer code comes to realize that programming largely revolves around *data*. Computer programs are all about receiving, manipulating, and returning data. Whether it's a simple program that calculates the sum of two numbers, or enterprise software that runs entire companies, software runs on data.

Data is a broad term that refers to all types of information, down to the most basic numbers and strings. In the simple but classic "Hello World!" program, the string "Hello World!" is a piece of data. In fact, even the most complex pieces of data usually break down into a bunch of numbers and strings.

Data structures refer to how data is organized. Let's look at the following code:

```
x = "Hello! "
y = "How are you "
z = "today?"
print x + y + z
```

This very simple program deals with three pieces of data, outputting three strings to make one coherent message. If we were to describe how the data is organized in this program, we'd say that we have three independent strings, each pointed to by a single variable.

You're going to learn in this book that the organization of data doesn't just matter for organization's sake, but can significantly impact *how fast your code runs*. Depending on how you choose to organize your data, your program may run faster or slower by orders of magnitude. And if you're building a program that needs to deal with lots of data, or a web app used by thousands of people simultaneously, the data structures you select may affect whether or not your software runs at all, or simply conks out because it can't handle the load.

When you have a solid grasp on the various data structures and each one's performance implications on the program that you're writing, you will have the keys to write fast and elegant code that will ensure that your software will run quickly and smoothly, and your expertise as a software engineer will be greatly enhanced.

In this chapter, we're going to begin our analysis of two data structures: arrays and sets. While the two data structures seem almost identical, you're going to learn the tools to analyze the performance implications of each choice.

The Array: The Foundational Data Structure

The *array* is one of the most basic data structures in computer science. We assume that you have worked with arrays before, so you are aware that an array is simply a list of data elements. The array is versatile, and can serve as a useful tool in many different situations, but let's just give one quick example.

If you are looking at the source code for an application that allows users to create and use shopping lists for the grocery store, you might find code like this:

```
array = ["apples", "bananas", "cucumbers", "dates", "elderberries"]
```

This array happens to contain five strings, each representing something that I might buy at the supermarket. (You've *got* to try elderberries.)

The *index* of an array is the number that identifies where a piece of data lives inside the array.

In most programming languages, we begin counting the index at 0. So for our example array, "apples" is at index 0, and "elderberries" is at index 4, like this:

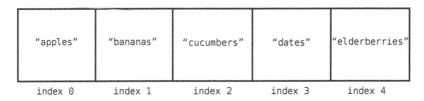

To understand the performance of a data structure—such as the array—we need to analyze the common ways that our code might interact with that data structure.

Most data structures are used in four basic ways, which we refer to as *operations*. They are:

- *Read:* Reading refers to looking something up from a particular spot within the data structure. With an array, this would mean looking up a value at a particular index. For example, looking up which grocery item is located at index 2 would be *reading* from the array.

- *Search:* Searching refers to looking for a particular value within a data structure. With an array, this would mean looking to see if a particular value exists within the array, and if so, which index it's at. For example, looking to see if "dates" is in our grocery list, and which index it's located at would be *searching* the array.

- *Insert:* Insertion refers to adding another value to our data structure. With an array, this would mean adding a new value to an additional slot within the array. If we were to add "figs" to our shopping list, we'd be *inserting* a new value into the array.

- *Delete:* Deletion refers to removing a value from our data structure. With an array, this would mean removing one of the values from the array. For example, if we removed "bananas" from our grocery list, that would be *deleting* from the array.

In this chapter, we'll analyze how fast each of these operations are when applied to an array.

And this brings us to the first Earth-shattering concept of this book: *when we measure how "fast" an operation takes, we do not refer to how fast the operation takes in terms of pure time, but instead in how many steps it takes.*

Why is this?

We can never say with definitiveness that any operation takes, say, five seconds. While the same operation may take five seconds on a particular computer, it may take longer on an older piece of hardware, or much faster on the supercomputers of tomorrow. Measuring the speed of an operation in terms of time is flaky, since it will always change depending on the hardware that it is run on.

However, we *can* measure the speed of an operation in terms of how many *steps* it takes. If Operation A takes five steps, and Operation B takes 500 steps, we can assume that Operation A will always be faster than Operation

B on *all* pieces of hardware. Measuring the number of steps is therefore the key to analyzing the speed of an operation.

Measuring the speed of an operation is also known as measuring its *time complexity*. Throughout this book, we'll use the terms *speed*, *time complexity*, *efficiency*, and *performance* interchangeably. They all refer to the number of steps that a given operation takes.

Let's jump into the four operations of an array and determine how many steps each one takes.

Reading

The first operation we'll look at is *reading*, which is looking up what value is contained at a particular index inside the array.

Reading from an array actually takes just one step. This is because the computer has the ability to jump to any particular index in the array and peer inside. In our example of ["apples", "bananas", "cucumbers", "dates", "elderberries"], if we looked up index 2, the computer would jump right to index 2 and report that it contains the value "cucumbers".

How is the computer able to look up an array's index in just one step? Let's see how:

A computer's memory can be viewed as a giant collection of cells. In the following diagram, you can see a grid of cells, in which some are empty, and some contain bits of data:

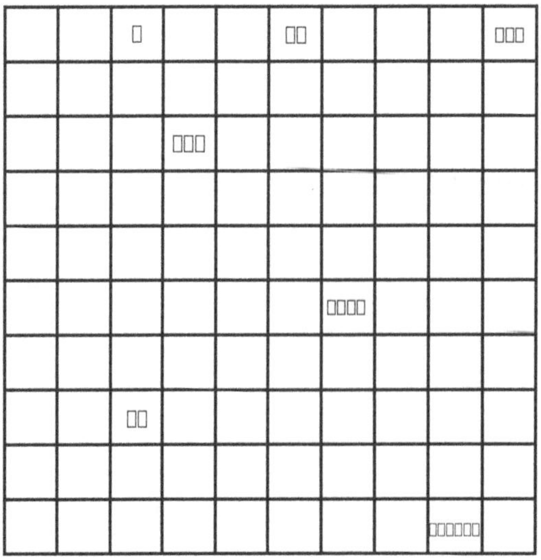

When a program declares an array, it allocates a contiguous set of empty cells for use in the program. So, if you were creating an array meant to hold five elements, your computer would find any group of five empty cells in a row and designate it to serve as your array:

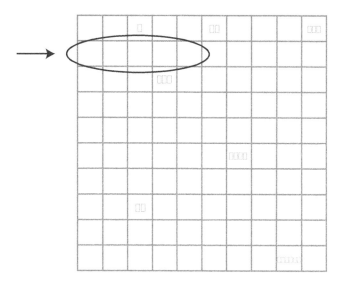

Now, every cell in a computer's memory has a specific address. It's sort of like a street address (for example, 123 Main St.), except that it's represented with a simple number. Each cell's memory address is one number greater than the previous cell. See the following diagram:

In the next diagram, we can see our shopping list array with its indexes and memory addresses:

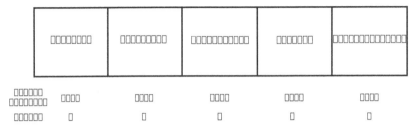

When the computer reads a value at a particular index of an array, it can jump straight to that index in one step because of the combination of the following facts:

1. A computer can jump to any memory address in one step. (Think of this as driving to 123 Main Street—you can drive there in one trip since you know exactly where it is.)

2. Recorded in each array is the memory address which it begins at. So the computer has this starting address readily.

3. Every array begins at index 0.

In our example, if we tell the computer to read the value at index 3, the computer goes through the following thought process:

1. Our array begins with index 0 at memory address 1010.

2. Index 3 will be exactly three slots past index 0.

3. So to find index 3, we'd go to memory address 1013, since 1010 + 3 is 1013.

Once the computer jumps to the memory address 1013, it returns the value, which is "dates".

Reading from an array is, therefore, a very efficient operation, since it takes just one step. An operation with just one step is naturally the fastest type of operation. One of the reasons that the array is such a powerful data structure is that we can look up the value at any index with such speed.

Now, what if instead of asking the computer what value is contained at index 3, we asked whether "dates" is contained within our array? That is the search operation, and we will explore that next.

Searching

As we stated previously, *searching* an array is looking to see whether a particular value exists within an array and if so, which index it's located at. Let's see how many steps the search operation takes for an array if we were to search for "dates".

When you and I look at the shopping list, our eyes immediately spot the "dates", and we can quickly count in our heads that it's at index 3. However, a computer doesn't have eyes, and needs to make its way through the array step by step.

To search for a value within an array, the computer starts at index 0, checks the value, and if it doesn't find what it's looking for, moves on to the next index. It does this until it finds the value it's seeking.

The following diagrams demonstrate this process when the computer searches for "dates" inside our grocery list array:

First, the computer checks index 0:

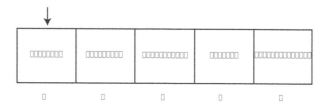

Since the value at index 0 is "apples", and not the "dates" that we're looking for, the computer moves on to the next index:

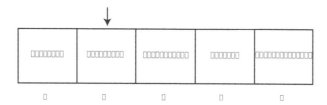

Since index 1 doesn't contain the "dates" we're looking for either, the computer moves on to index 2:

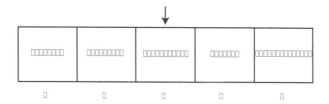

Once again, we're out of luck, so the computer moves to the next cell:

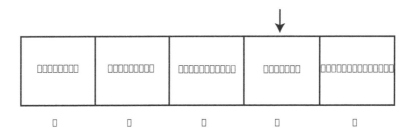

Aha! We've found the elusive "dates". We now know that the "dates" are at index 3. At this point, the computer does not need to move on to the next cell of the array, since we've already found what we're looking for.

In this example, since we had to check four different cells until we found the value we were searching for, we'd say that this particular operation took a total of four steps.

In *Why Algorithms Matter*, we'll learn about another way to search, but this basic search operation—in which the computer checks each cell one at a time—is known as *linear search*.

Now, what is the *maximum* number of steps a computer would need to conduct a linear search on an array?

If the value we're seeking happens to be in the final cell in the array (like "elderberries"), then the computer would end up searching through *every* cell of the array until it finally finds the value it's looking for. Also, if the value we're looking for doesn't occur in the array at all, the computer would likewise have to search every cell so it can be sure that the value doesn't exist within the array.

So it turns out that for an array of five cells, the maximum number of steps that linear search would take is five. For an array of 500 cells, the maximum number of steps that linear search would take is 500.

Another way of saying this is that for N cells in an array, linear search will take a maximum of N steps. In this context, N is just a variable that can be replaced by any number.

In any case, it's clear that searching is less efficient than reading, since searching can take many steps, while reading always takes just one step no matter how large the array.

Next, we'll analyze the operation of insertion—or, inserting a new value into an array.

Insertion

The efficiency of inserting a new piece of data inside an array depends on *where* inside the array you'd like to insert it.

Let's say we wanted to add "figs" to the very end of our shopping list. Such an insertion takes just one step. As we've seen earlier, the computer knows which memory address the array begins at. Now, the computer *also* knows how many elements the array currently contains, so it can calculate which memory address it needs to add the new element to, and can do so in one step. See the following diagram:

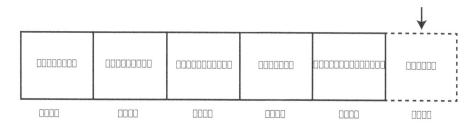

Inserting a new piece of data at the *beginning* or the *middle* of an array, however, is a different story. In these cases, we need to shift many pieces of data to make room for what we're inserting, leading to additional steps.

For example, let's say we wanted to add "figs" to index 2 within the array. See the following diagram:

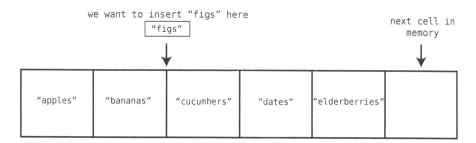

To do this, we need to move "cucumbers", "dates", and "elderberries" to the right to make room for the "figs". But even this takes multiple steps, since we need to first move "elderberries" one cell to the right to make room to move "dates". We then need to move "dates" to make room for the "cucumbers". Let's walk through this process.

Step #1: We move "elderberries" to the right:

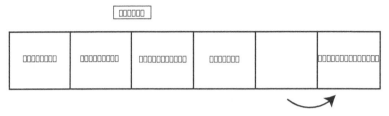

Step #2: We move "dates" to the right:

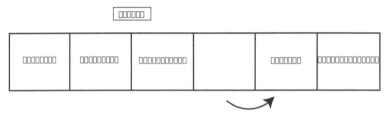

Step #3: We move "cucumbers" to the right:

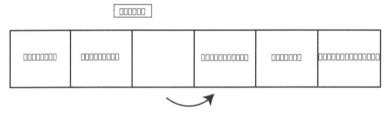

Step #4: Finally, we can insert "figs" into index 2:

We can see that in the preceding example, there were four steps. Three of the steps were shifting data to the right, while one step was the actual insertion of the new value.

The worst-case scenario for insertion into an array—that is, the scenario in which insertion takes the most steps—is where we insert data at the *beginning* of the array. This is because when inserting into the beginning of the array, we have to move *all* the other values one cell to the right.

So we can say that insertion in a worst-case scenario can take up to $N + 1$ *steps* for an array containing N elements. This is because the worst-case scenario is inserting a value into the beginning of the array in which there are N shifts (every data element of the array) and one insertion.

The final operation we'll explore—deletion—is like insertion, but in reverse.

Deletion

Deletion from an array is the process of eliminating the value at a particular index.

Let's return to our original example array, and delete the value at index 2. In our example, this would be the "cucumbers".

Step #1: We delete "cucumbers" from the array:

While the actual deletion of "cucumbers" technically took just one step, we now have a problem: we have an empty cell sitting smack in the middle of our array. An array is not allowed to have gaps in the middle of it, so to resolve this issue, we need to shift "dates" and "elderberries" to the left.

Step #2: We shift "dates" to the left:

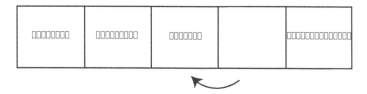

Step #3: We shift "elderberries" to the left:

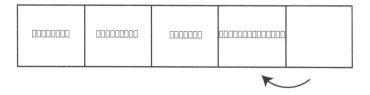

So it turns out that for this deletion, the entire operation took three steps. The first step was the actual deletion, and the other two steps were data shifts to close the gap.

So we've just seen that when it comes to deletion, the actual deletion itself is really just one step, but we need to follow up with additional steps of shifting data to the left to close the gap caused by the deletion.

Like insertion, the worst-case scenario of deleting an element is deleting the very first element of the array. This is because index 0 would be empty, which is not allowed for arrays, and we'd have to shift *all* the remaining elements to the left to fill the gap.

For an array of five elements, we'd spend one step deleting the first element, and four steps shifting the four remaining elements. For an array of 500 elements, we'd spend one step deleting the first element, and 499 steps shifting the remaining data. We can conclude, then, that for an array containing N elements, the maximum number of steps that deletion would take is N steps.

Now that we've learned how to analyze the time complexity of a data structure, we can discover how different data structures have different efficiencies. This is extremely important, since choosing the correct data structure for your program can have serious ramifications as to how performant your code will be.

The next data structure—the *set*—is so similar to the array that at first glance, it seems that they're basically the same thing. However, we'll see that the operations performed on arrays and sets have different efficiencies.

Sets: How a Single Rule Can Affect Efficiency

Let's explore another data structure: the *set*. A set is a data structure that does not allow duplicate values to be contained within it.

There are actually different types of sets, but for this discussion, we'll talk about an *array-based set*. This set is just like an array—it is a simple list of values. The only difference between this set and a classic array is that the set never allows duplicate values to be inserted into it.

For example, if you had the set ["a", "b", "c"] and tried to add another "b", the computer just wouldn't allow it, since a "b" already exists within the set.

Sets are useful when you need to ensure that you don't have duplicate data.

For instance, if you're creating an online phone book, you don't want the same phone number appearing twice. In fact, I'm currently suffering from this with my local phone book: my home phone number is not just listed for myself, but it is also erroneously listed as the phone number for some family named Zirkind. (Yes, this is a true story.) Let me tell you—it's quite annoying to receive phone calls and voicemails from people looking for the Zirkinds.

Sets: How a Single Rule Can Affect Efficiency • 13

For that matter, I'm sure the Zirkinds are also wondering why no one ever calls them. And when I call the Zirkinds to let them know about the error, my wife picks up the phone because I've called my own number. (Okay, that last part never happened.) If only the program that produced the phone book had used a set...

In any case, a set is an array with one simple constraint of not allowing duplicates. Yet, this constraint actually causes the set to have a *different efficiency* for one of the four primary operations.

Let's analyze the reading, searching, insertion, and deletion operations in context of an array-based set.

Reading from a set is exactly the same as reading from an array—it takes just one step for the computer to look up what is contained within a particular index. As we described earlier, this is because the computer can jump to any index within the set since it knows the memory address that the set begins at.

Searching a set also turns out to be no different than searching an array—it takes up to N steps to search to see if a value exists within a set. And deletion is also identical between a set and an array—it takes up to N steps to delete a value and move data to the left to close the gap.

Insertion, however, is where arrays and sets diverge. Let's first explore inserting a value at the *end* of a set, which was a best-case scenario for an array. With an array, the computer can insert a value at its end in a single step.

With a set, however, the computer *first needs to determine that this value doesn't already exist in this set*—because that's what sets do: they prevent duplicate data. So every insert *first requires a search*.

Let's say that our grocery list was a set—and that would be a decent choice since we don't want to buy the same thing twice, after all. If our current set is ["apples", "bananas", "cucumbers", "dates", "elderberries"], and we wanted to insert "figs", we'd have to conduct a search and go through the following steps:

Step #1: Search index 0 for "figs":

↓				
apples	bananas	cucumbers	dates	elderberries

It's not there, but it might be somewhere else in the set. We need to make sure that "figs" does not exist anywhere before we can insert it.

Step #2: Search index 1:

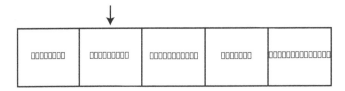

Step #3: Search index 2:

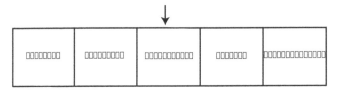

Step #4: Search index 3:

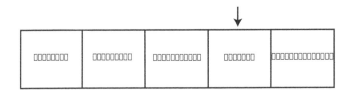

Step #5: Search index 4:

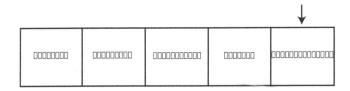

Only now that we've searched the entire set do we know that it's safe to insert "figs". And that brings us to our final step.

Step #6: Insert "figs" at the end of the set:

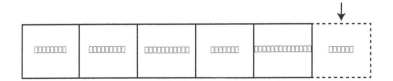

Inserting a value at the end of a set is still the best-case scenario, but we still had to perform six steps for a set originally containing five elements. That is, we had to search all five elements before performing the final insertion step.

Said another way: insertion into a set in a best-case scenario will take $N + 1$ steps for N elements. This is because there are N steps of search to ensure that the value doesn't already exist within the set, and then one step for the actual insertion.

In a worst-case scenario, where we're inserting a value at the *beginning* of a set, the computer needs to search N cells to ensure that the set doesn't already contain that value, and then another N steps to shift all the data to the right, and another final step to insert the new value. That's a total of $2N + 1$ steps.

Now, does this mean that you should avoid sets just because insertion is slower for sets than regular arrays? Absolutely not. Sets are important when you need to ensure that there is no duplicate data. (Hopefully, one day my phone book will be fixed.) But when you don't have such a need, an array may be preferable, since insertions for arrays are more efficient than insertions for sets. You must analyze the needs of your own application and decide which data structure is a better fit.

Wrapping Up

Analyzing the number of steps that an operation takes is the heart of understanding the performance of data structures. Choosing the right data structure for your program can spell the difference between bearing a heavy load vs. collapsing under it. In this chapter in particular, you've learned to use this analysis to weigh whether an array or a set might be the appropriate choice for a given application.

Now that we've begun to learn how to think about the time complexity of data structures, we can also use the same analysis to compare competing algorithms (even within the *same* data structure) to ensure the ultimate speed and performance of our code. And that's exactly what the next chapter is about.

CHAPTER 2

Why Algorithms Matter

In the previous chapter, we took a look at our first data structures and saw how choosing the right data structure can significantly affect the performance of our code. Even two data structures that seem so similar, such as the array and the set, can make or break a program if they encounter a heavy load.

In this chapter, we're going to discover that even if we decide on a particular data structure, there is another major factor that can affect the efficiency of our code: the proper selection of which *algorithm* to use.

Although the word *algorithm* sounds like something complex, it really isn't. An algorithm is simply a particular process for solving a problem. For example, the process for preparing a bowl of cereal can be called an algorithm. The cereal-preparation algorithm follows these four steps (for me, at least):

1. Grab a bowl.
2. Pour cereal in the bowl.
3. Pour milk in the bowl.
4. Dip a spoon in the bowl.

When applied to computing, an algorithm refers to a process for going about a particular operation. In the previous chapter, we analyzed four major operations, including reading, searching, insertion, and deletion. In this chapter, we'll see that at times, it's possible to go about an operation in more than one way. That is to say, there are multiple algorithms that can achieve a particular operation.

We're about to see how the selection of a particular algorithm can make our code either fast or slow—even to the point where it stops working under a lot of pressure. But first, let's take a look at a new data structure: the ordered array. We'll see how there is more than one algorithm for searching an ordered array, and we'll learn how to choose the right one.

Ordered Arrays

The *ordered array* is almost identical to the array we discussed in the previous chapter. The only difference is that ordered arrays require that the values are always kept—you guessed it—*in order*. That is, every time a value is added, it gets placed in the proper cell so that the values of the array remain sorted. In a standard array, on the other hand, values can be added to the end of the array without taking the order of the values into consideration.

For example, let's take the array [3, 17, 80, 202]:

Assume that we wanted to insert the value 75. If this array were a regular array, we could insert the 75 at the end, as follows:

As we noted in the previous chapter, the computer can accomplish this in a single step.

On the other hand, if this were an *ordered array*, we'd have no choice but to insert the 75 in the proper spot so that the values remained in ascending order:

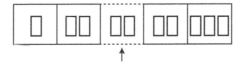

Now, this is easier said than done. The computer cannot simply drop the 75 into the right slot in a single step, because first it has to *find* the right place in which the 75 needs to be inserted, and then shift other values to make room for it. Let's break down this process step by step.

Let's start again with our original ordered array:

Step #1: We check the value at index 0 to determine whether the value we want to insert—the 75—should go to its left or to its right:

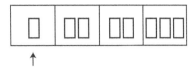

Since 75 is greater than 3, we know that the 75 will be inserted somewhere to its right. However, we don't know yet exactly which cell it should be inserted into, so we need to check the next cell.

Step #2: We inspect the value at the next cell:

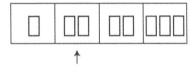

75 is greater than 17, so we need to move on.

Step #3: We check the value at the next cell:

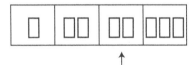

We've encountered the value 80, which is *greater* than the 75 that we wish to insert. Since we've reached the first value which is greater than 75, we can conclude that the 75 must be placed immediately to the left of this 80 to maintain the order of this ordered array. To do that, we need to shift data to make room for the 75.

Step #4: Move the final value to the right:

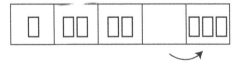

Step #5: Move the next-to-last value to the right:

Step #6: We can finally insert the 75 into its correct spot:

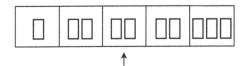

It emerges that when inserting into an ordered array, we need to always conduct a search before the actual insertion to determine the correct spot for the insertion. That is one key difference (in terms of efficiency) between a standard array and an ordered array.

While insertion is less efficient by an ordered array than in a regular array, the ordered array has a secret superpower when it comes to the search operation.

Searching an Ordered Array

In the previous chapter, we described the process for searching for a particular value within a regular array: we check each cell one at a time—from left to right—until we find the value we're looking for. We noted that this process is referred to as linear search.

Let's see how linear search differs between a regular and ordered array.

Say that we have a regular array of [17, 3, 75, 202, 80]. If we were to search for the value 22—which happens to be nonexistent in our example—we would need to search each and every element because the 22 could potentially be anywhere in the array. The only time we could stop our search before we reach the array's end is if we happen to find the value we're looking for before we reach the end.

With an ordered array, however, we can stop a search early even if the value isn't contained within the array. Let's say we're searching for a 22 within an ordered array of [3, 17, 75, 80, 202]. We can stop the search as soon as we reach the 75, since it's impossible that the 22 is anywhere to the right of it.

Here's a Ruby implementation of linear search on an ordered array:

```ruby
def linear_search(array, value)
  # We iterate through every element in the array:
  array.each do |element|
    # If we find the value we're looking for, we return it:
    if element == value
      return value
```

```
    # If we reach an element that is greater than the value
    # we're looking for, we can exit the loop early:
    elsif element > value
      break
    end
  end

  # We return nil if we do not find the value within the array:
  return nil
end
```

In this light, linear search will take fewer steps in an ordered array vs. a standard array in most situations. That being said, if we're searching for a value that happens to be the final value or greater than the final value, we will still end up searching each and every cell.

At first glance, then, standard arrays and ordered arrays don't have tremendous differences in efficiency.

But that is because we have not yet unlocked the power of algorithms. And that is about to change.

We've been assuming until now that the only way to search for a value within an ordered array is linear search. The truth, however, is that linear search is only *one possible algorithm*—that is, it is one particular process for going about searching for a value. It is the process of searching each and every cell until we find the desired value. But it is not the *only* algorithm we can use to search for a value.

The big advantage of an ordered array over a regular array is that an ordered array allows for an alternative searching algorithm. This algorithm is known as *binary search*, and it is a much, *much* faster algorithm than linear search.

Binary Search

You've probably played this guessing game when you were a child (or maybe you play it with your children now): I'm thinking of a number between 1 and 100. Keep on guessing which number I'm thinking of, and I'll let you know whether you need to guess higher or lower.

You know intuitively how to play this game. You wouldn't start the guessing by choosing the number 1. You'd start with 50 which is smack in the middle. Why? Because by selecting 50, no matter whether I tell you to guess higher or lower, you've automatically eliminated half the possible numbers!

If you guess 50 and I tell you to guess higher, you'd then pick 75, to eliminate half of the *remaining* numbers. If after guessing 75, I told you to guess lower, you'd pick 62 or 63. You'd keep on choosing the halfway mark in order to keep eliminating half of the remaining numbers.

Let's visualize this process with a similar game, except that we're told to guess a number between 1 and 10:

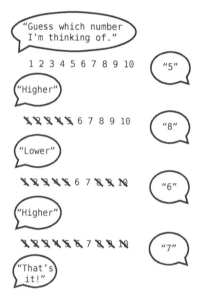

This, in a nutshell, is binary search.

The major advantage of an ordered array over a standard array is that we have the option of performing a binary search rather than a linear search. Binary search is impossible with a standard array because the values can be in any order.

To see this in action, let's say we had an ordered array containing nine elements. To the computer, it doesn't know what value each cell contains, so we will portray the array like this:

Say that we'd like to search for the value 7 inside this ordered array. We can do so using binary search:

Step #1: We begin our search from the central cell. We can easily jump to this cell since we know the length of the array, and can just divide that number by two and jump to the appropriate memory address. We check the value at that cell:

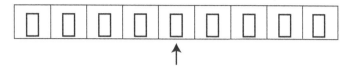

Since the value uncovered is a 9, we can conclude that the 7 is somewhere to its left. We've just successfully eliminated half of the array's cells—that is, all the cells to the right of the 9 (and the 9 itself):

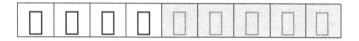

Step #2: Among the cells to the left of the 9, we inspect the middlemost value. There are two middlemost values, so we arbitrarily choose the left one:

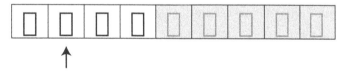

It's a 4, so the 7 must be somewhere to its right. We can eliminate the 4 and the cell to its left:

Step #3: There are two more cells where the 7 can be. We arbitrarily choose the left one:

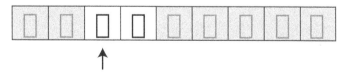

Step #4: We inspect the final remaining cell. (If it's not there, that means that there is no 7 within this ordered array.)

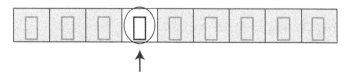

We've found the 7 successfully in four steps. While this is the same number of steps linear search would have taken in this example, we'll demonstrate the power of binary search shortly.

Here's an implementation of binary search in Ruby:

```ruby
def binary_search(array, value)

  # First, we establish the lower and upper bounds of where the value
  # we're searching for can be. To start, the lower bound is the first
  # value in the array, while the upper bound is the last value:

  lower_bound = 0
  upper_bound = array.length - 1

  # We begin a loop in which we keep inspecting the middlemost value
  # between the upper and lower bounds:

  while lower_bound <= upper_bound do

    # We find the midpoint between the upper and lower bounds:
    # (We don't have to worry about the result being a non-integer
    # since in Ruby, the result of division of integers will always
    # be rounded down to the nearest integer.)

    midpoint = (upper_bound + lower_bound) / 2

    # We inspect the value at the midpoint:

    value_at_midpoint = array[midpoint]

    # If the value at the midpoint is the one we're looking for, we're done.
    # If not, we change the lower or upper bound based on whether we need
    # to guess higher or lower:

    if value < value_at_midpoint
      upper_bound = midpoint - 1
    elsif value > value_at_midpoint
      lower_bound = midpoint + 1
    elsif value == value_at_midpoint
      return midpoint
    end
  end

  # If we've narrowed the bounds until they've reached each other, that
  # means that the value we're searching for is not contained within
  # this array:

  return nil
end
```

Binary Search vs. Linear Search

With ordered arrays of a small size, the algorithm of binary search doesn't have much of an advantage over the algorithm of linear search. But let's see what happens with larger arrays.

With an array containing one hundred values, here are the maximum numbers of steps it would take for each type of search:

- Linear search: one hundred steps
- Binary search: seven steps

With linear search, if the value we're searching for is in the final cell or is greater than the value in the final cell, we have to inspect each and every element. For an array the size of 100, this would take one hundred steps.

When we use binary search, however, each guess we make eliminates half of the possible cells we'd have to search. In our very first guess, we get to eliminate a whopping fifty cells.

Let's look at this another way, and we'll see a pattern emerge:

With an array of size 3, the maximum number of steps it would take to find something using binary search is two.

If we double the number of cells in the array (and add one more to keep the number odd for simplicity's sake), there are seven cells. For such an array, the maximum number of steps to find something using binary search is three.

If we double it again (and add one) so that the ordered array contains fifteen elements, the maximum number of steps to find something using binary search is four.

The pattern that emerges is that for every time we double the number of items in the ordered array, the number of steps needed for binary search increases by just one.

This pattern is unusually efficient: for every time we double the data, the binary search algorithm adds a maximum of just one more step.

Contrast this with linear search. If you had three items, you'd need up to three steps. For seven elements, you'd need a maximum of seven steps. For one hundred, you'd need up to one hundred steps. With linear search, *there are as many steps as there are items*. For linear search, every time we double the number of elements in the array, we *double* the number of steps we need to find something. For binary search, every time we double the number of elements in the array, we only need to add *one more step*.

We can visualize the difference in performance between linear and binary search with this graph on page 26.

Let's see how this plays out for even larger arrays. With an array of 10,000 elements, a linear search can take up to 10,000 steps, while binary search

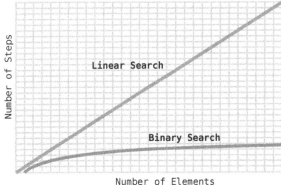

takes up to a maximum of just thirteen steps. For an array the size of one million, linear search would take up to one million steps, while binary search would take up to just *twenty steps.*

Again, keep in mind that ordered arrays aren't faster in every respect. As you've seen, insertion in ordered arrays is slower than in standard arrays. But here's the trade-off: by using an ordered array, you have somewhat slower insertion, but much faster search. Again, you must always analyze your application to see what is a better fit.

Wrapping Up

And this is what algorithms are all about. Often, there is more than one way to achieve a particular computing goal, and the algorithm you choose can seriously affect the speed of your code.

It's also important to realize that there usually isn't a single data structure or algorithm that is perfect for every situation. For example, just because ordered arrays allow for binary search doesn't mean you should always use ordered arrays. In situations where you don't anticipate searching the data much, but only adding data, standard arrays may be a better choice because their insertion is faster.

As we've seen, the way to analyze competing algorithms is to count the number of steps each one takes.

In the next chapter, we're going to learn about a formalized way of expressing the time complexity of competing data structures and algorithms. Having this common language will give us clearer information that will allow us to make better decisions about which algorithms we choose.

CHAPTER 3

Oh Yes! Big O Notation

We've seen in the preceding chapters that the number of steps that an algorithm takes is the primary factor in determining its efficiency.

However, we can't simply label one algorithm a "22-step algorithm" and another a "400-step algorithm." This is because the number of steps that an algorithm takes cannot be pinned down to a single number. Let's take linear search, for example. The number of steps that linear search takes varies, as it takes as many steps as there are cells in the array. If the array contains twenty-two elements, linear search takes twenty-two steps. If the array has 400 elements, however, linear search takes 400 steps.

The more accurate way to quantify efficiency of linear search is to say that linear search takes *N steps* for *N elements in the array*. Of course, that's a pretty wordy way of expressing this concept.

In order to help ease communication regarding time complexity, computer scientists have borrowed a concept from the world of mathematics to describe a concise and consistent language around the efficiency of data structures and algorithms. Known as *Big O Notation*, this formalized expression around these concepts allows us to easily categorize the efficiency of a given algorithm and convey it to others.

Once you understand Big O Notation, you'll have the tools to analyze every algorithm going forward in a consistent and concise way—and it's the way that the pros use.

While Big O Notation comes from the math world, we're going to leave out all the mathematical jargon and explain it as it relates to computer science. Additionally, we're going to begin by explaining Big O Notation in very simple terms, and continue to refine it as we proceed through this and the next three chapters. It's not a difficult concept, but we'll make it even easier by explaining it in chunks over multiple chapters.

Big O: Count the Steps

Instead of focusing on units of time, Big O achieves consistency by focusing only on the *number of steps* that an algorithm takes.

In *Why Data Structures Matter*, we discovered that reading from an array takes just one step, no matter how large the array is. The way to express this in Big O Notation is:

O(1)

Many pronounce this verbally as "Big Oh of 1." Others call it "Order of 1." My personal preference is "Oh of 1." While there is no standardized way to *pronounce* Big O Notation, there is only one way to *write* it.

O(1) simply means that the algorithm takes the same number of steps no matter how much data there is. In this case, reading from an array always takes just one step no matter how much data the array contains. On an old computer, that step may have taken twenty minutes, and on today's hardware it may take just a nanosecond. But in both cases, the algorithm takes just a single step.

Other operations that fall under the category of O(1) are the insertion and deletion of a value at the end of an array. As we've seen, each of these operations takes just one step for arrays of any size, so we'd describe their efficiency as O(1).

Let's examine how Big O Notation would describe the efficiency of linear search. Recall that linear search is the process of searching an array for a particular value by checking each cell, one at a time. In a worst-case scenario, linear search will take as many steps as there are elements in the array. As we've previously phrased it: for N elements in the array, linear search can take up to a maximum of N steps.

The appropriate way to express this in Big O Notation is:

O(N)

I pronounce this as "Oh of N."

O(N) is the "Big O" way of saying that for N elements inside an array, the algorithm would take N steps to complete. It's that simple.

> ### So Where's the Math?
>
> As I mentioned, in this book, I'm taking an easy-to-understand approach to the topic of Big O. That's not the only way to do it; if you were to take a traditional college course on algorithms, you'd probably be introduced to Big O from a mathematical perspective. Big O is originally a concept from mathematics, and therefore it's often described in mathematical terms. For example, one way of describing Big O is that it describes the upper bound of the growth rate of a function, or that if a function g(x) grows no faster than a function f(x), then g is said to be a member of O(f). Depending on your mathematics background, that either makes sense, or doesn't help very much. I've written this book so that you don't need as much math to understand the concept.
>
> If you want to dig further into the math behind Big O, check out *Introduction to Algorithms* by Thomas H. Cormen, Charles E. Leiserson, Ronald L. Rivest, and Clifford Stein (MIT Press, 2009) for a full mathematical explanation. Justin Abrahms provides a pretty good definition in his article: https://justin.abrah.ms/computer-science/understanding-big-o-formal-definition.html. Also, the Wikipedia article on Big O (https://en.wikipedia.org/wiki/Big_O_notation) takes a fairly heavy mathematical approach.

Constant Time vs. Linear Time

Now that we've encountered O(N), we can begin to see that Big O Notation does more than simply describe the number of steps that an algorithm takes, such as a hard number such as 22 or 400. Rather, it describes how many steps an algorithm takes *based on the number of data elements that the algorithm is acting upon*. Another way of saying this is that Big O answers the following question: *how does the number of steps change as the data increases?*

An algorithm that is O(N) will take as many steps as there are elements of data. So when an array increases in size by one element, an O(N) algorithm will increase by one step. An algorithm that is O(1) will take the same number of steps no matter how large the array gets.

Look at how these two types of algorithms are plotted on a graph on page 30.

You'll see that O(N) makes a perfect diagonal line. This is because for every additional piece of data, the algorithm takes one additional step. Accordingly, the more data, the more steps the algorithm will take. For the record, O(N) is also known as *linear time*.

Contrast this with O(1), which is a perfect horizontal line, since the number of steps in the algorithm remains constant no matter how much data there is. Because of this, O(1) is also referred to as *constant time*.

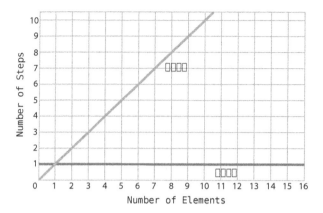

As Big O is primarily concerned about how an algorithm performs across varying amounts of data, an important point emerges: an algorithm can be described as O(1) even if it takes more than one step. Let's say that a particular algorithm always takes *three* steps, rather than one—but it always takes these three steps no matter how much data there is. On a graph, such an algorithm would look like this:

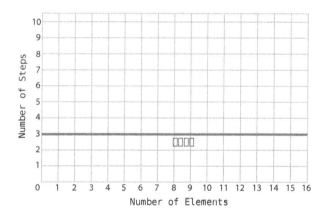

Because the number of steps remains constant no matter how much data there is, this would also be considered constant time and be described by Big O Notation as O(1). Even though the algorithm technically takes three steps rather than one step, Big O Notation considers that trivial. O(1) is the way to describe *any* algorithm that doesn't change its number of steps even when the data increases.

If a three-step algorithm is considered O(1) as long as it remains constant, it follows that even a constant 100-step algorithm would be expressed as O(1) as well. While a 100-step algorithm is less efficient than a one-step algorithm, the fact that it is O(1) still makes it *more efficient* than any O(N) algorithm.

Why is this?

See the following graph:

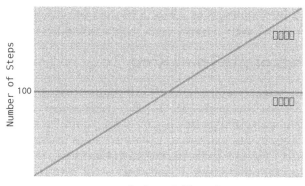

As the graph depicts, for an array of fewer than one hundred elements, O(N) algorithm takes fewer steps than the O(1) 100-step algorithm. At exactly one hundred elements, the two algorithms take the same number of steps (100). But here's the key point: for *all arrays greater than one hundred*, the O(N) algorithm takes more steps.

Because there will always be *some* amount of data in which the tides turn, and O(N) takes more steps from that point until infinity, O(N) is considered to be, on the whole, less efficient than O(1).

The same is true for an O(1) algorithm that always takes one million steps. As the data increases, there will inevitably reach a point where O(N) becomes less efficient than the O(1) algorithm, and will remain so up until an infinite amount of data.

Same Algorithm, Different Scenarios

As we learned in the previous chapters, linear search isn't *always* O(N). It's true that if the item we're looking for is in the final cell of the array, it will take N steps to find it. But where the item we're searching for is found in the *first* cell of the array, linear search will find the item in just one step. Technically, this would be described as O(1). If we were to describe the efficiency of linear search in its totality, we'd say that linear search is O(1) in a *best-case* scenario, and O(N) in a *worst-case* scenario.

While Big O effectively describes both the best- and worst-case scenarios of a given algorithm, Big O Notation generally refers to *worst-case scenario* unless specified otherwise. This is why most references will describe linear search as being O(N) even though it *can* be O(1) in a best-case scenario.

The reason for this is that this "pessimistic" approach can be a useful tool: knowing exactly how inefficient an algorithm can get in a worst-case scenario prepares us for the worst and may have a strong impact on our choices.

An Algorithm of the Third Kind

In the previous chapter, we learned that binary search on an ordered array is much faster than linear search on the same array. Let's learn how to describe binary search in terms of Big O Notation.

We can't describe binary search as being O(1), because the number of steps increases as the data increases. It also doesn't fit into the category of O(N), since the number of steps is much fewer than the number of elements that it searches. As we've seen, binary search takes only seven steps for an array containing one hundred elements.

Binary search seems to fall somewhere *in between* O(1) and O(N).

In Big O, we describe binary search as having a time complexity of:

O(log N)

I pronounce this as "Oh of log N." This type of algorithm is also known as having a time complexity of *log time*.

Simply put, O(log N) is the Big O way of describing an algorithm that *increases one step each time the data is doubled*. As we learned in the previous chapter, binary search does just that. We'll see momentarily why this is expressed as O(log N), but let's first summarize what we've learned so far.

Of the three types of algorithms we've learned about so far, they can be sorted from most efficient to least efficient as follows:

O(1)

O(log N)

O(N)

Let's look at a graph on page 33 that compares the three types.

Note how O(log N) curves ever-so-slightly upwards, making it less efficient than O(1), but much more efficient than O(N).

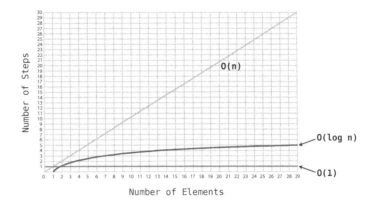

To understand why this algorithm is called "O(log N)," we need to first understand what *logarithms* are. If you're already familiar with this mathematical concept, you can skip the next section.

Logarithms

Let's examine why algorithms such as binary search are described as O(log N). What is a log, anyway?

Log is shorthand for *logarithm*. The first thing to note is that logarithms have nothing to do with algorithms, even though the two words look and sound so similar.

Logarithms are the inverse of *exponents*. Here's a quick refresher on what exponents are:

2^3 is the equivalent of:

2 * 2 * 2

which just happens to be 8.

Now, $\log_2 8$ is the converse of the above. It means: how many times do you have to multiply 2 by itself to get a result of 8?

Since you have to multiply 2 by itself 3 times to get 8, $\log_2 8 = 3$.

Here's another example:

2^6 translates to:

2 * 2 * 2 * 2 * 2 * 2 = 64

Since, we had to multiply 2 by itself 6 times to get 64,

$\log_2 64 = 6$.

While the preceding explanation is the official "textbook" definition of logarithms, I like to use an alternative way of describing the same concept because many people find that they can wrap their heads around it more easily, especially when it comes to Big O Notation.

Another way of explaining $\log_2 8$ is: if we kept *dividing* 8 by 2 until we ended up with 1, how many 2s would we have in our equation?

8 / 2 / 2 / 2 = 1

In other words, how many times do we need to divide 8 by 2 until we end up with 1? In this example, it takes us 3 times. Therefore,

$\log_2 8 = 3$.

Similarly, we could explain $\log_2 64$ as: how many times do we need to halve 64 until we end up with 1?

64 / 2 / 2 / 2 / 2 / 2 / 2 = 1

Since there are 6 2s, $\log_2 64 = 6$.

Now that you understand what logarithms are, the meaning behind O(log N) will become clear.

O(log N) Explained

Let's bring this all back to Big O Notation. Whenever we say O(log N), it's actually shorthand for saying $O(\log_2 N)$. We're just omitting that small 2 for convenience.

Recall that O(N) means that for N data elements, the algorithm would take N steps. If there are eight elements, the algorithm would take eight steps.

O(log N) means that for N data elements, the algorithm would take *$\log_2 N$ steps*. If there are eight elements, the algorithm would take three steps, since $\log_2 8 = 3$.

Said another way, if we keep dividing the eight elements in half, it would take us three steps until we end up with one element.

This is *exactly* what happens with binary search. As we search for a particular item, we keep dividing the array's cells in half until we narrow it down to the correct number.

Said simply: *O(log N) means that the algorithm takes as many steps as it takes to keep halving the data elements until we remain with one.*

The following table demonstrates a striking difference between the efficiencies of O(N) and O(log N):

N Elements	O(N)	O(log N)
8	8	3
16	16	4
32	32	5
64	64	6
128	128	7
256	256	8
512	512	9
1024	1024	10

While the O(N) algorithm takes as many steps as there are data elements, the O(log N) algorithm takes just one additional step every time the data elements are doubled.

In future chapters, we will encounter algorithms that fall under categories of Big O Notation other than the three we've learned about so far. But in the meantime, let's apply these concepts to some examples of everyday code.

Practical Examples

Here's some typical Python code that prints out all the items from a list:

```
things = ['apples', 'baboons', 'cribs', 'dulcimers']
for thing in things:
    print "Here's a thing: %s" % thing
```

How would we describe the efficiency of this algorithm in Big O Notation?

The first thing to realize is that this is an example of an algorithm. While it may not be fancy, any code that does anything at all is technically an algorithm—it's a particular process for solving a problem. In this case, the problem is that we want to print out all the items from a list. The algorithm we use to solve this problem is a for loop containing a print statement.

To break this down, we need to analyze how many steps this algorithm takes. In this case, the main part of the algorithm—the for loop—takes four steps. In this example, there are four things in the list, and we print each one out one time.

However, this process isn't constant. If the list would contain ten elements, the for loop would take ten steps. Since this for loop takes as many steps as there are elements, we'd say that this algorithm has an efficiency of O(N).

Let's take another example. Here is one of the most basic code snippets known to mankind:

```
print 'Hello world!'
```

The time complexity of the preceding algorithm (printing "Hello world!") is O(1), because it always takes one step.

The next example is a simple Python-based algorithm for determining whether a number is prime:

```
def is_prime(number):
    for i in range(2, number):
        if number % i == 0:
            return False
    return True
```

The preceding code accepts a number as an argument and begins a for loop in which we divide every number from 2 up to that number and see if there's a remainder. If there's no remainder, we know that the number is not prime and we immediately return False. If we make it all the way up to the number and always find a remainder, then we know that the number is prime and we return True.

The efficiency of this algorithm is O(N). In this example, the data does not take the form of an array, but the actual number passed in as an argument. If we pass the number 7 into is_prime, the for loop runs for about seven steps. (It really runs for five steps, since it starts at two and ends right before the actual number.) For the number 101, the loop runs for about 101 steps. Since the number of steps increases in lockstep with the number passed into the function, this is a classic example of O(N).

Wrapping Up

Now that we understand Big O Notation, we have a consistent system that allows us to compare any two algorithms. With it, we will be able to examine real-life scenarios and choose between competing data structures and algorithms to make our code faster and able to handle heavier loads.

In the next chapter, we'll encounter a real-life example in which we use Big O Notation to speed up our code significantly.

CHAPTER 4

Speeding Up Your Code with Big O

Big O Notation is a great tool for comparing competing algorithms, as it gives an objective way to measure them. We've already been able to use it to quantify the difference between binary search vs. linear search, as binary search is O(log N)—a much faster algorithm than linear search, which is O(N).

However, there may not always be two clear alternatives when writing everyday code. Like most programmers, you probably use whatever approach pops into your head first. With Big O, you have the opportunity to compare your algorithm to *general algorithms out there in the world*, and you can say to yourself, "Is this a fast or slow algorithm as far as algorithms generally go?"

If you find that Big O labels your algorithm as a "slow" one, you can now take a step back and try to figure out if there's a way to optimize it by trying to get it to fall under a faster category of Big O. This may not always be possible, of course, but it's certainly worth thinking about before concluding that it's not.

In this chapter, we'll write some code to solve a practical problem, and then measure our algorithm using Big O. We'll then see if we might be able to modify the algorithm in order to give it a nice efficiency bump. (Spoiler: we will.)

Bubble Sort

Before jumping into our practical problem, though, we need to first learn about a new category of algorithmic efficiency in the world of Big O. To demonstrate it, we'll get to use one of the classic algorithms of computer science lore.

Sorting algorithms have been the subject of extensive research in computer science, and tens of such algorithms have been developed over the years. They all solve the following problem:

Given an array of unsorted numbers, how can we sort them so that they end up in ascending order?

In this and the following chapters, we're going to encounter a number of these sorting algorithms. Some of the first ones we'll be learning about are known as "simple sorts," in that they are easier to understand, but are not as efficient as some of the faster sorting algorithms out there.

Bubble Sort is a very basic sorting algorithm, and follows these steps:

1. Point to two consecutive items in the array. (Initially, we start at the very beginning of the array and point to its first two items.) Compare the first item with the second one:

   | 2 | 1 | 3 | 5 |

2. If the two items are out of order (in other words, the left value is greater than the right value), swap them:

   | 2 | 1 | 3 | 5 |

   | 1 | 2 | 3 | 5 |

   (If they already happen to be in the correct order, do nothing for this step.)

3. Move the "pointers" one cell to the right:

   | 1 | 2 | 3 | 5 |

   Repeat steps 1 and 2 until we reach the end of the array or any items that have already been sorted.

4. Repeat steps 1 through 3 until we have a round in which we didn't have to make any swaps. This means that the array is in order.

   Each time we repeat steps 1 through 3 is known as a *passthrough*. That is, we "passed through" the primary steps of the algorithm, and will repeat the same process until the array is fully sorted.

Bubble Sort in Action

Let's walk through a complete example. Assume that we wanted to sort the array [4, 2, 7, 1, 3]. It's currently out of order, and we want to produce an array containing the same values in the correct, ascending order.

Let's begin Passthrough #1:

This is our starting array:

$$\boxed{4}\boxed{2}\boxed{7}\boxed{1}\boxed{3}$$

Step #1: First, we compare the 4 and the 2. They're out of order:

$$\boxed{4}\boxed{2}\boxed{7}\boxed{1}\boxed{3}$$

Step #2: So we swap them:

$$\boxed{4}\boxed{2}\boxed{7}\boxed{1}\boxed{3}$$

$$\boxed{2}\boxed{4}\boxed{7}\boxed{1}\boxed{3}$$

Step #3: Next, we compare the 4 and the 7:

$$\boxed{2}\boxed{4}\boxed{7}\boxed{1}\boxed{3}$$

They're in the correct order, so we don't need to perform any swaps.

Step #4: We now compare the 7 and the 1:

$$\boxed{2}\boxed{4}\boxed{7}\boxed{1}\boxed{3}$$

Step #5: They're out of order, so we swap them:

$$\boxed{2}\boxed{4}\boxed{7}\boxed{1}\boxed{3}$$

$$\boxed{2}\boxed{4}\boxed{1}\boxed{7}\boxed{3}$$

Step #6: We compare the 7 and the 3:

$$\boxed{2}\boxed{4}\boxed{1}\boxed{7}\boxed{3}$$

Step #7: They're out of order, so we swap them:

$$2\;4\;1\;7\;3$$

$$2\;4\;1\;3\;7$$

We now know for a fact that the 7 is in its correct position within the array, because we kept moving it along to the right until it reached its proper place. We've put little lines surrounding it to indicate this fact.

This is actually the reason that this algorithm is called *Bubble* Sort: in each passthrough, the highest unsorted value "bubbles" up to its correct position.

Since we made at least one swap during this passthrough, we need to conduct another one.

We begin Passthrough #2:

The 7 is already in the correct position:

$$2\;4\;1\;3\;7$$

Step #8: We begin by comparing the 2 and the 4:

$$2\;4\;1\;3\;7$$

They're in the correct order, so we can move on.

Step #9: We compare the 4 and the 1:

$$2\;4\;1\;3\;7$$

Step #10: They're out of order, so we swap them:

$$2\;4\;1\;3\;7$$

$$2\;1\;4\;3\;7$$

Step #11: We compare the 4 and the 3:

```
2 1 4 3 7
    ↑ ↑
```

Step #12: They're out of order, so we swap them:

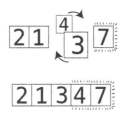

```
2 1 3 4 7
```

We don't have to compare the 4 and the 7 because we know that the 7 is already in its correct position from the previous passthrough. And now we also know that the 4 is bubbled up to its correct position as well. This concludes our second passthrough. Since we made at least one swap during this passthrough, we need to conduct another one.

We now begin Passthrough #3:

Step #13: We compare the 2 and the 1:

```
2 1 3 4 7
↑ ↑
```

Step #14: They're out of order, so we swap them:

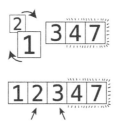

```
1 2 3 4 7
  ↑ ↑
```

Step #15: We compare the 2 and the 3:

```
1 2 3 4 7
```

They're in the correct order, so we don't need to swap them.

We now know that the 3 has bubbled up to its correct spot. Since we made at least one swap during this passthrough, we need to perform another one.

And so begins Passthrough #4:

Step #16: We compare the 1 and the 2:

Since they're in order, we don't need to swap. We can end this passthrough, since all the remaining values are already correctly sorted.

Now that we've made a passthrough that didn't require any swaps, we know that our array is completely sorted:

Bubble Sort Implemented

Here's an implementation of Bubble Sort in Python:

```python
def bubble_sort(list):
    unsorted_until_index = len(list) - 1
    sorted = False

    while not sorted:
        sorted = True
        for i in range(unsorted_until_index):
            if list[i] > list[i+1]:
                sorted = False
                list[i], list[i+1] = list[i+1], list[i]
        unsorted_until_index = unsorted_until_index - 1

list = [65, 55, 45, 35, 25, 15, 10]
bubble_sort(list)
print list
```

Let's break this down line by line. We'll first present the line of code, followed by its explanation.

```
unsorted_until_index = len(list) - 1
```

We keep track of up to which index is still unsorted with the unsorted_until_index variable. At the beginning, the array is totally unsorted, so we initialize this variable to be the final index in the array.

```
sorted = False
```

We also create a sorted variable that will allow us to keep track whether the array is fully sorted. Of course, when our code first runs, it isn't.

```
while not sorted:
    sorted = True
```

We begin a while loop that will last as long as the array is not sorted. Next, we preliminarily establish sorted to be True. We'll change this back to False as soon as we have to make any swaps. If we get through an entire passthrough without having to make any swaps, we'll know that the array is completely sorted.

```
for i in range(unsorted_until_index):
    if list[i] > list[i+1]:
        sorted = False
        list[i], list[i+1] = list[i+1], list[i]
```

Within the while loop, we begin a for loop that starts from the beginning of the array and goes until the index that has not yet been sorted. Within this loop, we compare every pair of adjacent values, and swap them if they're out of order. We also change sorted to False if we have to make a swap.

```
unsorted_until_index = unsorted_until_index - 1
```

By this line of code, we've completed another passthrough, and can safely assume that the value we've bubbled up to the right is now in its correct position. Because of this, we decrement the unsorted_until_index by 1, since the index it was already pointing to is now sorted.

Each round of the while loop represents another passthrough, and we run it until we know that our array is fully sorted.

The Efficiency of Bubble Sort

The Bubble Sort algorithm contains two kinds of steps:

- *Comparisons*: two numbers are compared with one another to determine which is greater.
- *Swaps*: two numbers are swapped with one another in order to sort them.

Let's start by determining how many *comparisons* take place in Bubble Sort.

Our example array has five elements. Looking back, you can see that in our first passthrough, we had to make four comparisons between sets of two numbers.

In our second passthrough, we had to make only three comparisons. This is because we didn't have to compare the final two numbers, since we knew that the final number was in the correct spot due to the first passthrough.

In our third passthrough, we made two comparisons, and in our fourth passthrough, we made just one comparison.

So, that's:

4 + 3 + 2 + 1 = 10 comparisons.

To put it more generally, we'd say that for N elements, we make

(N - 1) + (N - 2) + (N - 3) ... + 1 comparisons.

Now that we've analyzed the number of comparisons that take place in Bubble Sort, let's analyze the *swaps*.

In a worst-case scenario, where the array is not just randomly shuffled, but sorted in descending order (the exact opposite of what we want), we'd actually need a swap for each comparison. So we'd have ten comparisons and ten swaps in such a scenario for a grand total of twenty steps.

So let's look at the complete picture. With an array containing ten elements in reverse order, we'd have:

9 + 8 + 7 + 6 + 5 + 4 + 3 + 2 + 1 = 45 comparisons, and another forty-five swaps. That's a total of ninety steps.

With an array containing *twenty* elements, we'd have:

19 + 18 + 17 + 16 + 15 + 14 + 13 + 12 + 11 + 10 + 9 + 8 + 7 + 6 + 5 + 4 + 3 + 2 + 1 = *190* comparisons, and approximately 190 swaps, for a total of 380 steps.

Notice the inefficiency here. As the number of elements increase, the number of steps grows exponentially. We can see this clearly with the following table:

N data elements	Max # of steps
5	20
10	90
20	380
40	1560
80	6320

If you look precisely at the growth of steps as N increases, you'll see that it's growing by approximately N^2.

N data elements	# of Bubble Sort steps	N^2
5	20	25
10	90	100
20	380	400
40	1560	1600
80	6320	6400

Therefore, in Big O Notation, we would say that Bubble Sort has an efficiency of $O(N^2)$.

Said more officially: in an $O(N^2)$ algorithm, for N data elements, there are roughly N^2 steps.

$O(N^2)$ is considered to be a relatively inefficient algorithm, since as the data increases, the steps increase dramatically. Look at this graph:

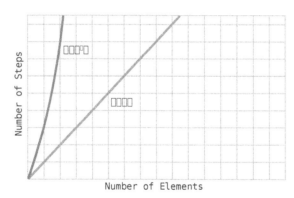

Note how $O(N^2)$ curves sharply upwards in terms of number of steps as the data grows. Compare this with $O(N)$, which plots along a simple, diagonal line.

One last note: $O(N^2)$ is also referred to as *quadratic time*.

A Quadratic Problem

Let's say you're writing a JavaScript application that requires you to check whether an array contains any duplicate values.

One of the first approaches that may come to mind is the use of nested for loops, as follows:

```
function hasDuplicateValue(array) {
    for(var i = 0; i < array.length; i++) {
        for(var j = 0; j < array.length; j++) {
            if(i !== j && array[i] == array[j]) {
                return true;
            }
        }
    }
    return false;
}
```

In this function, we iterate through each element of the array using var i. As we focus on each element in i, we then run a *second* for loop that checks through all of the elements in the array—using var j—and checks if the elements at positions i and j are the same. If they are, that means we've encountered duplicates. If we get through all of the looping and we haven't encountered any duplicates, we can return false, indicating that there are no duplicates in the given array.

While this certainly works, is it efficient? Now that we know a bit about Big O Notation, let's take a step back and see what Big O would say about this function.

Remember that Big O is a measure of how many steps our algorithm would take relative to how much data there is. To apply that to our situation, we'd ask ourselves: for N elements in the array provided to our hasDuplicateValue function, how many steps would our algorithm take in a worst-case scenario?

To answer the preceding question, we first need to determine what qualifies as a step as well as what the worst-case scenario would be.

The preceding function has one type of step, namely *comparisons*. It repeatedly compares i and j to see if they are equal and therefore represent a duplicate. In a worst-case scenario, the array contains no duplicates, which would force our code to complete all of the loops and exhaust every possible comparison before returning false.

Based on this, we can conclude that for N elements in the array, our function would perform N^2 comparisons. This is because we perform an outer loop that must iterate N times to get through the entire array, and for *each iteration*, we must iterate *another N times* with our inner loop. That's N steps * N steps, which is otherwise known as N^2 steps, leaving us with an algorithm of $O(N^2)$.

We can actually prove this by adding some code to our function that tracks the number of steps:

```
function hasDuplicateValue(array) {
    var steps = 0;
    for(var i = 0; i < array.length; i++) {
        for(var j = 0; j < array.length; j++) {
            steps++;
            if(i !== j && array[i] == array[j]) {
                return true;
            }
        }
    }
    console.log(steps);
    return false;
}
```

If we run hasDuplicateValue([1,2,3]), we'll see an output of 9 in the JavaScript console, indicating that there were nine comparisons. Since there are nine steps for an array of three elements, this is a classic example of $O(N^2)$.

Unsurprisingly, $O(N^2)$ is the efficiency of algorithms in which nested loops are used. When you see a nested loop, $O(N^2)$ alarm bells should start going off in your head.

While our implementation of the hasDuplicateValue function is the only algorithm we've developed so far, the fact that it's $O(N^2)$ should give us pause. This is because $O(N^2)$ is considered a relatively slow algorithm. Whenever encountering a slow algorithm, it's worth spending some time to think about whether there may be any faster alternatives. This is especially true if you anticipate that your function may need to handle large amounts of data, and your application may come to a screeching halt if not optimized properly. There may *not* be any better alternatives, but let's first make sure.

A Linear Solution

Below is another implementation of the hasDuplicateValue function that doesn't rely upon nested loops. Let's analyze it and see if it's any more efficient than our first implementation.

```
function hasDuplicateValue(array) {
    var existingNumbers = [];
    for(var i = 0; i < array.length; i++) {
        if(existingNumbers[array[i]] === undefined) {
            existingNumbers[array[i]] = 1;
        } else {
            return true;
        }
    }
    return false;
}
```

This implementation uses a single loop, and keeps track of which numbers it has already encountered using an array called existingNumbers. It uses this array in an interesting way: every time the code encounters a new number, it stores the value 1 inside the index of the existingNumbers corresponding to that number.

For example, if the given array is [3, 5, 8], by the time the loop is finished, the existingNumbers array will look like this:

[undefined, undefined, undefined, 1, undefined, 1, undefined, undefined, 1]

The 1s are located in indexes 3, 5, and 8 to indicate that those numbers are found in the given array.

However, before the code stores 1 in the appropriate index, it first checks to see whether that index already has 1 as its value. If it does, that means we already encountered that number before, meaning that we've found a duplicate.

To determine the efficiency of this new algorithm in terms of Big O, we once again need to determine the number of steps the algorithm takes in a worst-case scenario. Like our first implementation, the significant type of step that our algorithm has is comparisons. That is, it looks up a particular index of the existingNumbers array and compares it to undefined, as expressed in this code:

```
if(existingNumbers[array[i]] === undefined)
```

(In addition to the comparisons, we do also make *insertions* into the existingNumbers array, but we're considering that trivial in this analysis. More on this in the next chapter.)

Once again, the worst-case scenario is when the array contains no duplicates, in which case our function must complete the entire loop.

This new algorithm appears to make N comparisons for N data elements. This is because there's only one loop, and it simply iterates for as many elements as there are in the array. We can test out this theory by tracking the steps in the JavaScript console:

```
function hasDuplicateValue(array) {
    var steps = 0;
    var existingNumbers = [];
    for(var i = 0; i < array.length; i++) {
        steps++;
        if(existingNumbers[array[i]] === undefined) {
            existingNumbers[array[i]] = 1;
        } else {
            return true;
        }
    }
```

```
        console.log(steps);
        return false;
}
```

When running hasDuplicateValue([1,2,3]), we see that the output in the JavaScript console is 3, which is the same number of elements in our array.

Using Big O Notation, we'd conclude that this new implementation is O(N).

We know that O(N) is much faster than O(N^2), so by using this second approach, we've optimized our hasDuplicateValue significantly. If our program will handle lots of data, this will make a *big* difference. (There is actually one disadvantage with this new implementation, but we won't discuss it until the final chapter.)

Wrapping Up

It's clear that having a solid understanding of Big O Notation can allow us to identify slow code and select the faster of two competing algorithms. However, there are situations in which Big O Notation will have us believe that two algorithms have the same speed, while one is actually faster. In the next chapter, we're going to learn how to evaluate the efficiencies of various algorithms even when Big O isn't nuanced enough to do so.

CHAPTER 5

Optimizing Code with and Without Big O

We've seen that Big O is a great tool for contrasting algorithms and determining which algorithm should be used for a given situation. However, it's certainly not the *only* tool. In fact, there are times where two competing algorithms may be described in exactly the same way using Big O Notation, yet one algorithm is significantly faster than the other.

In this chapter, we're going to learn how to discern between two algorithms that *seem* to have the same efficiency, and select the faster of the two.

Selection Sort

In the previous chapter, we explored an algorithm for sorting data known as Bubble Sort, which had an efficiency of $O(N^2)$. We're now going to dig into another sorting algorithm called *Selection Sort*, and see how it measures up to Bubble Sort.

The steps of Selection Sort are as follows:

1. We check each cell of the array from left to right to determine which value is least. As we move from cell to cell, we keep in a variable the lowest value we've encountered so far. (Really, we keep track of its index, but for the purposes of the following diagrams, we'll just focus on the actual value.) If we encounter a cell that contains a value that is even less than the one in our variable, we replace it so that the variable now points to the new index. See the following diagram:

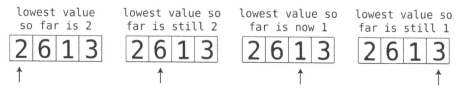

2. Once we've determined which index contains the lowest value, we swap that index with the value we began the passthrough with. This would be index 0 in the first passthrough, index 1 in the second passthrough, and so on and so forth. In the next diagram, we make the swap of the first passthrough:

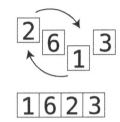

3. Repeat steps 1 and 2 until all the data is sorted.

Selection Sort in Action

Using the example array [4, 2, 7, 1, 3], our steps would be as follows:

We begin our first passthrough:

We set things up by inspecting the value at index 0. By definition, it's the lowest value in the array that we've encountered so far (as it's the *only* value we've encountered so far), so we keep track of its index in a variable:

lowest value so far is 4

| 4 | 2 | 7 | 1 | 3 |
↑

Step #1: We compare the 2 with the lowest value so far (which happens to be 4):

lowest value = 4

| 4 | 2 | 7 | 1 | 3 |
  ↑

The 2 is even less than the 4, so it becomes the lowest value so far:

lowest value = 2

| 4 | 2 | 7 | 1 | 3 |
  ↑

Step #2: We compare the next value—the 7—with the lowest value so far. The 7 is greater than the 2, so 2 remains our lowest value:

lowest value = 2

| 4 | 2 | 7 | 1 | 3 |
    ↑

Step #3: We compare the 1 with the lowest value so far:

lowest value = 2

| 4 | 2 | 7 | 1 | 3 |

↑

Since the 1 is even less than the 2, the 1 becomes our new lowest value:

lowest value = 1

| 4 | 2 | 7 | 1 | 3 |

↑

Step #4: We compare the 3 to the lowest value so far, which is the 1. We've reached the end of the array, and we've determined that 1 is the lowest value out of the entire array:

lowest value = 1

| 4 | 2 | 7 | 1 | 3 |

↑

Step #5: Since 1 is the lowest value, we swap it with whatever value is at index 0—the index we began this passthrough with:

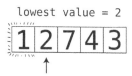

We have now determined that the 1 is in its correct place within the array:

| 1 | 2 | 7 | 4 | 3 |

We are now ready to begin our second passthrough:

Setup: the first cell—index 0—is already sorted, so this passthrough begins at the next cell, which is index 1. The value at index 1 is the number 2, and it is the lowest value we've encountered in this passthrough so far:

lowest value = 2

| 1 | 2 | 7 | 4 | 3 |

↑

Step #6: We compare the 7 with the lowest value so far. The 2 is less than the 7, so the 2 remains our lowest value:

Step #7: We compare the 4 with the lowest value so far. The 2 is less than the 4, so the 2 remains our lowest value:

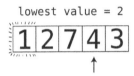

Step #8: We compare the 3 with the lowest value so far. The 2 is less than the 3, so the 2 remains our lowest value:

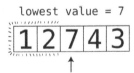

We've reached the end of the array. We don't need to perform any swaps in this passthrough, and we can therefore conclude that the 2 is in its correct spot. This ends our second passthrough, leaving us with:

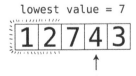

We now begin Passthrough #3:

Setup: we begin at index 2, which contains the value 7. The 7 is the lowest value we've encountered so far in this passthrough:

Step #9: We compare the 4 with the 7:

We note that 4 is our new lowest value:

```
            lowest value = 4
            ┌─┬─┬─┬─┬─┐
            │1│2│7│4│3│
            └─┴─┴─┴─┴─┘
                   ↑
```

Step #10: We encounter the 3, which is even lower than the 4:

```
            lowest value = 4
            ┌─┬─┬─┬─┬─┐
            │1│2│7│4│3│
            └─┴─┴─┴─┴─┘
                     ↑
```

3 is now our new lowest value:

```
            lowest value = 3
            ┌─┬─┬─┬─┬─┐
            │1│2│7│4│3│
            └─┴─┴─┴─┴─┘
                     ↑
```

Step #11: We've reached the end of the array, so we swap the 3 with the value that we started our passthrough at, which is the 7:

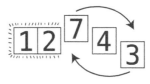

We now know that the 3 is in the correct place within the array:

```
            ┌─┬─┬─┬─┬─┐
            │1│2│3│4│7│
            └─┴─┴─┴─┴─┘
```

While you and I can both see that the entire array is correctly sorted at this point, the *computer* does not know this yet, so it must begin a fourth passthrough:

Setup: we begin the passthrough with index 3. The 4 is the lowest value so far:

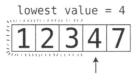

Step #12: We compare the 7 with the 4:

The 4 remains the lowest value we've encountered in this passthrough so far, so we don't need any swaps and we know it's in the correct place.

Since all the cells besides the last one are correctly sorted, that must mean that the last cell is also in the correct order, and our entire array is properly sorted:

Selection Sort Implemented

Here's a JavaScript implementation of Selection Sort:

```javascript
function selectionSort(array) {
  for(var i = 0; i < array.length; i++) {
    var lowestNumberIndex = i;
    for(var j = i + 1; j < array.length; j++) {
      if(array[j] < array[lowestNumberIndex]) {
        lowestNumberIndex = j;
      }
    }

    if(lowestNumberIndex != i) {
      var temp = array[i];
      array[i] = array[lowestNumberIndex];
      array[lowestNumberIndex] = temp;
    }
  }
  return array;
}
```

Let's break this down. We'll first present the line of code, followed by its explanation.

```javascript
for(var i = 0; i < array.length; i++) {
```

Here, we have an outer loop that represents each passthrough of Selection Sort. We then begin keeping track of the *index* containing the lowest value we encounter so far with:

```javascript
var lowestNumberIndex = i;
```

which sets lowestNumberIndex to be whatever index i represents. Note that we're actually tracking the index of the lowest number instead of the actual number itself. This index will be 0 at the beginning of the first passthrough, 1 at the beginning of the second, and so on.

```
for(var j = i + 1; j < array.length; j++) {
```

kicks off an inner for loop that starts at i + 1.

```
if(array[j] < array[lowestNumberIndex]) {
  lowestNumberIndex = j;
}
```

checks each element of the array that has not yet been sorted and looks for the lowest number. It does this by keeping track of the index of the lowest number it found so far in the lowestNumberIndex variable.

By the end of the inner loop, we've determined the index of the lowest number not yet sorted.

```
if(lowestNumberIndex != i) {
  var temp = array[i];
  array[i] = array[lowestNumberIndex];
  array[lowestNumberIndex] = temp;
}
```

We then check to see if this lowest number is already in its correct place (i). If not, we swap the lowest number with the number that's in the position that the lowest number should be at.

The Efficiency of Selection Sort

Selection Sort contains two types of steps: comparisons and swaps. That is, we compare each element with the lowest number we've encountered in each passthrough, and we swap the lowest number into its correct position.

Looking back at our example of an array containing five elements, we had to make a total of ten comparisons. Let's break it down.

Passthrough #	# of comparisons
1	4 comparisons
2	3 comparisons
3	2 comparisons
4	1 comparison

So that's a grand total of 4 + 3 + 2 + 1 = 10 comparisons.

To put it more generally, we'd say that for N elements, we make

(N - 1) + (N - 2) + (N - 3) ... + 1 comparisons.

As for *swaps*, however, we only need to make a maximum of one swap per passthrough. This is because in each passthrough, we make either one or zero swaps, depending on whether the lowest number of that passthrough is already in the correct position. Contrast this with Bubble Sort, where in a worst-case scenario—an array in descending order—we have to make a swap for *each and every* comparison.

Here's the side-by-side comparison between Bubble Sort and Selection Sort:

N elements	Max # of steps in Bubble Sort	Max # of steps in Selection Sort
5	20	14 (10 comparisons + 4 swaps)
10	90	54 (45 comparisons + 9 swaps)
20	380	199 (180 comparisons + 19 swaps)
40	1560	819 (780 comparisons + 39 swaps)
80	6320	3239 (3160 comparisons + 79 swaps)

From this comparison, it's clear that Selection Sort contains about half the number of steps that Bubble Sort does, indicating that Selection Sort is twice as fast.

Ignoring Constants

But here's the funny thing: in the world of Big O Notation, Selection Sort and Bubble Sort are described in exactly the same way.

Again, Big O Notation describes how many steps are required relative to the number of data elements. So it would seem at first glance that the number of steps in Selection Sort are roughly *half of whatever N^2 is*. It would therefore seem reasonable that we'd describe the efficiency of Selection Sort as being $O(N^2 / 2)$. That is, for N data elements, there are $N^2 / 2$ steps. The following table bears this out:

N elements	$N^2 / 2$	Max # of steps in Selection Sort
5	$5^2 / 2 = 12.5$	14
10	$10^2 / 2 = 50$	54
20	$20^2 / 2 = 200$	199
40	$40^2 / 2 = 800$	819
80	$80^2 / 2 = 3200$	3239

In reality, however, Selection Sort is described in Big O as $O(N^2)$, just like Bubble Sort. This is because of a major rule of Big O that we're now introducing for the first time:

Big O Notation ignores constants.

This is simply a mathematical way of saying that Big O Notation never includes regular numbers that aren't an exponent.

In our case, what should technically be $O(N^2 / 2)$ becomes simply $O(N^2)$. Similarly, $O(2N)$ would become $O(N)$, and $O(N / 2)$ would also become $O(N)$. Even $O(100N)$, which is *100 times slower than O(N)*, would also be referred to as $O(N)$.

Offhand, it would seem that this rule would render Big O Notation entirely useless, as you can have two algorithms that are described in the same exact way with Big O, and yet one can be *100 times faster* than the other. And that's exactly what we're seeing here with Selection Sort and Bubble Sort. Both are described in Big O as $O(N^2)$, but Selection Sort is actually twice as fast as Bubble Sort. And indeed, if given the choice between these two options, Selection Sort is the better choice.

So, what gives?

The Role of Big O

Despite the fact that Big O doesn't distinguish between Bubble Sort and Selection Sort, it is still very important, because it serves as a great way to classify the *long-term growth rate* of algorithms. That is, for *some amount of data*, $O(N)$ will be always be faster than $O(N^2)$. And this is true no matter whether the $O(N)$ is really $O(2N)$ or even $O(100N)$ under the hood. It is a fact that there is *some amount* of data at which even $O(100N)$ will become faster than $O(N^2)$. (We've seen essentially the same concept in *Oh Yes! Big O Notation* when comparing a 100-step algorithm with $O(N)$, but we'll reiterate it here in our current context.)

Look at the first graph on page 60, in which we compare $O(N)$ with $O(N^2)$.

We've seen this graph in the previous chapter. It depicts how $O(N)$ is faster than $O(N^2)$ for *all* amounts of data.

Now take a look at the second graph on page 60, where we compare $O(100N)$ with $O(N^2)$.

In this second graph, we see that $O(N^2)$ is faster than $O(100N)$ for certain amounts of data, but after a point, even $O(100N)$ becomes faster and remains faster for all increasing amounts of data from that point onward.

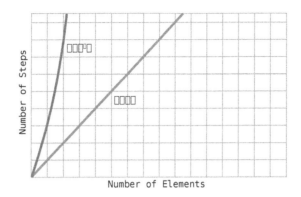

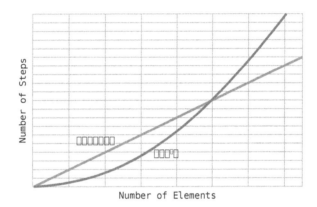

It is for this very reason that Big O ignores constants. The purpose of Big O is that for different classifications, *there will be a point* at which one classification supersedes the other in speed, and will remain faster forever. When that point occurs exactly, however, is not the concern of Big O.

Because of this, there really is no such thing as O(100N)—it is simply written as O(N).

Similarly, with large amounts of data, O(log N) will always be faster than O(N), even if the given O(log N) algorithm is actually O(2 * log N) under the hood.

So Big O is an extremely useful tool, because if two algorithms fall under different classifications of Big O, you'll generally know which algorithm to use since with large amounts of data, one algorithm is guaranteed to be faster than the other at a certain point.

However, the main takeaway of this chapter is that when two algorithms fall under the *same* classification of Big O, it doesn't necessarily mean that both algorithms process at the same speed. After all, Bubble Sort is twice as slow as Selection Sort even though both are O(N^2). So while Big O is perfect for contrasting algorithms that fall under different classifications of Big O, when two algorithms fall under the *same* classification, further analysis is required to determine which algorithm is faster.

A Practical Example

Let's say you're tasked with writing a Ruby application that takes an array and creates a new array out of *every other element* from the original array. It might be tempting to use the each_with_index method available to arrays to loop through the original array as follows:

```ruby
def every_other(array)
  new_array = []

  array.each_with_index do |element, index|
    new_array << element if index.even?
  end

  return new_array
end
```

In this implementation, we iterate through each element of the original array and only add the element to the new array if the index of that element is an even number.

When analyzing the steps taking place here, we can see that there are really two types of steps. We have one type of step in which we look up each element of the array, and another type of step in which we add elements to the new array.

We perform N array lookups, since we loop through each and every element of the array. We only perform N / 2 insertions, though, since we only insert every other element into the new array. Since we have N lookups, and we have N / 2 insertions, we would say that our algorithm technically has an efficiency of O(N + (N / 2)), which we can also rephrase as O(1.5N). But since Big O Notation throws out the constants, we would say that our algorithm is simply O(N).

While this algorithm does work, we always want to take a step back and see if there's room for any optimization. And in fact, we can.

Instead of iterating through each element of the array and checking whether the index is an even number, we can instead simply look up every other element of the array in the first place:

```
def every_other(array)
  new_array = []
  index = 0

  while index < array.length
    new_array << array[index]
    index += 2
  end

  return new_array
end
```

In this second implementation, we use a while loop to skip over each element, rather than check each one. It turns out that for N elements, there are N / 2 lookups and N / 2 insertions into the new array. Like the first implementation, we'd say that the algorithm is O(N).

However, our first implementation truly takes 1.5N steps, while our second implementation only takes N steps, making our second implementation significantly faster. While the first implementation is more idiomatic in the way Ruby programmers write their code, if we're dealing with large amounts of data, it's worth considering using the second implementation to get a significant performance boost.

Wrapping Up

We now have some very powerful analysis tools at our disposal. We can use Big O to determine broadly how efficient an algorithm is, and we can also compare two algorithms that fall within one classification of Big O.

However, there is another important factor to take into account when comparing the efficiencies of two algorithms. Until now, we've focused on how slow an algorithm is in a worst-case scenario. Now, worst-case scenarios, by definition, don't happen all the time. On average, the scenarios that occur are—well—average-case scenarios. In the next chapter, we'll learn how to take all scenarios into account.

CHAPTER 6

Optimizing for Optimistic Scenarios

When evaluating the efficiency of an algorithm, we've focused primarily on how many steps the algorithm would take in a worst-case scenario. The rationale behind this is simple: if you're prepared for the worst, things will turn out okay.

However, we'll discover in this chapter that the worst-case scenario isn't the *only* situation worth considering. Being able to consider *all* scenarios is an important skill that can help you choose the appropriate algorithm for every situation.

Insertion Sort

We've previously encountered two different sorting algorithms: Bubble Sort and Selection Sort. Both have efficiencies of $O(N^2)$, but Selection Sort is actually twice as fast. Now we'll learn about a third sorting algorithm called *Insertion Sort* that will reveal the power of analyzing scenarios beyond the worst case.

Insertion Sort consists of the following steps:

1. In the first passthrough, we temporarily remove the value at index 1 (the second cell) and store it in a temporary variable. This will leave a gap at that index, since it contains no value:

    | 8 | 4 | 2 | 3 |

    | 4 |
    | 8 | ↑ | 2 | 3 |

    In subsequent passthroughs, we remove the values at the subsequent indexes.

2. We then begin a shifting phase, where we take each value to the left of the gap, and compare it to the value in the temporary variable:

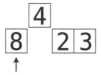

If the value to the left of the gap is greater than the temporary variable, we shift that value to the right:

As we shift values to the right, inherently, the gap moves leftwards. As soon as we encounter a value that is lower than the temporarily removed value, or we reach the left end of the array, this shifting phase is over.

3. We then insert the temporarily removed value into the current gap:

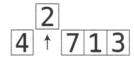

4. We repeat steps 1 through 3 until the array is fully sorted.

Insertion Sort in Action

Let's apply Insertion Sort to the array: [4, 2, 7, 1, 3].

We begin the first passthrough by inspecting the value at index 1. This happens to contain the value 2:

Setup: we temporarily remove the 2, and keep it inside a variable called temp_value. We represent this value by shifting it above the rest of the array:

Step #1: We compare the 4 to the temp_value, which is 2:

```
  2
4   7 1 3
↑
```

Step #2: Since 4 is greater than 2, we shift the 4 to the right:

```
2
4 7 1 3
↘
```

There's nothing left to shift, as the gap is now at the left end of the array.

Step #3: We insert the temp_value back into the array, completing our first passthrough:

```
↙
2 4 7 1 3
```

We begin Passthrough #2:

Setup: in our second passthrough, we temporarily remove the value at index 2. We'll store this in temp_value. In this case, the temp_value is 7:

```
    7
2 4 ↑ 1 3
```

Step #4: We compare the 4 to the temp_value:

```
    7
2 4   1 3
  ↑
```

4 is lower, so we won't shift it. Since we reached a value that is less than the temp_value, this shifting phase is over.

Step #5: We insert the temp_value back into the gap, ending the second passthrough:

$$\downarrow$$
$$\boxed{2}\boxed{4}\boxed{7}\boxed{1}\boxed{3}$$

Passthrough #3:

Setup: we temporarily remove the 1, and store it in temp_value:

$$\boxed{1}$$
$$\boxed{2}\boxed{4}\boxed{7}\uparrow\boxed{3}$$

Step #6: We compare the 7 to the temp_value:

$$\boxed{1}$$
$$\boxed{2}\boxed{4}\boxed{7}\ \ \boxed{3}$$
$$\uparrow$$

Step #7: 7 is greater than 1, so we shift the 7 to the right:

$$\boxed{1}$$
$$\boxed{2}\boxed{4}\ \ \boxed{7}\boxed{3}$$

Step #8: We compare the 4 to the temp_value:

$$\boxed{1}$$
$$\boxed{2}\boxed{4}\ \ \boxed{7}\boxed{3}$$
$$\uparrow$$

Step #9: 4 is greater than 1, so we shift it as well:

$$\boxed{1}$$
$$\boxed{2}\ \ \boxed{4}\boxed{7}\boxed{3}$$

Step #10: We compare the 2 to the temp_value:

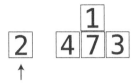

Step #11: The 2 is greater, so we shift it:

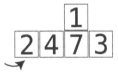

Step #12: The gap has reached the left end of the array, so we insert the temp_value into the gap, concluding this passthrough:

```
1 2 4 7 3
```

Passthrough #4:

Setup: we temporarily remove the value from index 4, making it our temp_value. This is the value 3:

```
        3
1 2 4 7
```

Step #13: We compare the 7 to the temp_value:

```
        3
1 2 4 7
      ↑
```

Step #14: The 7 is greater, so we shift the 7 to the right:

```
        3
1 2 4   7
```

Step #15: We compare the 4 to the temp_value:

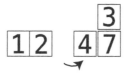

Step #16: The 4 is greater than the 3, so we shift the 4:

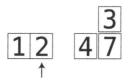

Step #17: We compare the 2 to the temp_value. 2 is less than 3, so our shifting phase is complete:

Step #18: We insert the temp_value back into the gap:

Our array is now fully sorted:

Insertion Sort Implemented

Here is a Python implementation of Insertion Sort:

```python
def insertion_sort(array):
    for index in range(1, len(array)):

        position = index
        temp_value = array[index]

        while position > 0 and array[position - 1] > temp_value:
            array[position] = array[position - 1]
            position = position - 1

        array[position] = temp_value
```

Let's walk through this step by step. We'll first present the line of code, followed by its explanation.

```
for index in range(1, len(array)):
```

First, we start a loop beginning at index 1 that runs through the entire array. The current index is kept in the variable index.

```
position = index
temp_value = array[index]
```

Next, we mark a position at whatever index currently is. We also assign the value at that index to the variable temp_value.

```
while position > 0 and array[position - 1] > temp_value:
    array[position] = array[position - 1]
    position = position - 1
```

We then begin an inner while loop. We check whether the value to the left of position is greater than the temp_value. If it is, we then use array[position] = array[position - 1] to shift that left value one cell to the right, and then decrement position by one. We then check whether the value to the left of the new position is greater than temp_value, and keep repeating this process until we find a value that is less than the temp_value.

```
array[position] = temp_value
```

Finally, we drop the temp_value into the gap within the array.

The Efficiency of Insertion Sort

There are four types of steps that occur in Insertion Sort: removals, comparisons, shifts, and insertions. To analyze the efficiency of Insertion Sort, we need to tally up each of these steps.

First, let's dig into *comparisons*. A comparison takes place each time we compare a value to the left of the gap with the temp_value.

In a worst-case scenario, where the array is sorted in reverse order, we have to compare every number to the left of temp_value with temp_value in each passthrough. This is because each value to the left of temp_value will always be greater than temp_value, so the passthrough will only end when the gap reaches the left end of the array.

During the first passthrough, in which temp_value is the value at index 1, a maximum of one comparison is made, since there is only one value to the left of the temp_value. On the second passthrough, a maximum of two comparisons are made, and so on and so forth. On the final passthrough, we need to compare the temp_value with every single value in the array besides temp_value itself. In other words, if there are N elements in the array, a maximum of N - 1 comparisons are made in the final passthrough.

We can, therefore, formulate the total number of comparisons as:

1 + 2 + 3 + ... + N - 1 comparisons.

In our example of an array containing five elements, that's a maximum of:

1 + 2 + 3 + 4 = 10 comparisons.

For an array containing ten elements, there would be:

1 + 2 + 3 + 4 + 5 + 6 + 7 + 8 + 9 = 45 comparisons.

(For an array containing twenty elements, there would be a total of 190 comparisons, and so on.)

When examining this pattern, it emerges that for an array containing N elements, there are *approximately* $N^2 / 2$ comparisons. ($10^2 / 2$ is 50, and $20^2 / 2$ is 200.)

Let's continue analyzing the other types of steps.

Shifts occur each time we move a value one cell to the right. When an array is sorted in reverse order, there will be as many shifts as there are comparisons, since every comparison will force us to shift a value to the right.

Let's add up comparisons and shifts for a worst-case scenario:

$N^2 / 2$ comparisons

+ $N^2 / 2$ shifts

_____

N^2 steps

Removing and inserting the temp_value from the array happen once per passthrough. Since there are always N - 1 passthroughs, we can conclude that there are N - 1 removals and N - 1 insertions.

So now we've got:

N^2 comparisons & shifts combined

N - 1 removals

+ N - 1 insertions

_____

$N^2 + 2N - 2$ steps

We've already learned one major rule of Big O—that Big O ignores constants. With this rule in mind, we'd—at first glance—simplify this to $O(N^2 + N)$.

However, there is another major rule of Big O that we'll reveal now:

Big O Notation only takes into account the highest order of N.

That is, if we have some algorithm that takes $N^4 + N^3 + N^2 + N$ steps, we only consider N^4 to be significant—and just call it $O(N^4)$. Why is this?

Look at the following table:

N	N^2	N^3	N^4
2	4	8	16
5	25	125	625
10	100	1,000	10,000
100	10,000	1,000,000	100,000,000
1,000	1,000,000	1,000,000,000	1,000,000,000,000

As N increases, N^4 becomes so much more significant than any other order of N. When N is 1,000, N^4 is 1,000 times greater than N^3. Because of this, we only consider the greatest order of N.

In our example, then, $O(N^2 + N)$ simply becomes $O(N^2)$.

It emerges that in a worst-case scenario, Insertion Sort has the same time complexity as Bubble Sort and Selection Sort. They're all $O(N^2)$.

We noted in the previous chapter that although Bubble Sort and Selection Sort are both $O(N^2)$, Selection Sort is faster since Selection Sort has $N^2 / 2$ steps compared with Bubble Sort's N^2 steps. At first glance, then, we'd say that Insertion Sort is as slow as Bubble Sort, since it too has N^2 steps. (It's really $N^2 + 2N - 2$ steps.)

If we'd stop the book here, you'd walk away thinking that Selection Sort is the best choice out of the three, since it's twice as fast as either Bubble Sort or Insertion Sort. But it's actually not that simple.

The Average Case

Indeed, in a worst-case scenario, Selection Sort *is* faster than Insertion Sort. However, it is critical that we also take into account the *average-case scenario*.

Why?

By definition, the cases that occur most frequently are average scenarios. The worst- and best-case scenarios happen only rarely. Let's look at this simple bell curve:

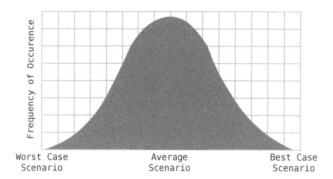

Best- and worst-case scenarios happen relatively infrequently. In the real world, however, average scenarios are what occur most of the time.

And this makes a lot of sense. Think of a randomly sorted array. What are the odds that the values will occur in perfect ascending or descending order? It's much more likely that the values will be all over the place.

Let's examine Insertion Sort in context of all scenarios.

We've looked at how Insertion Sort performs in a worst-case scenario—where the array sorted in descending order. In the worst case, we pointed out that in each passthrough, we compare and shift every value that we encounter. (We calculated this to be a total of N^2 comparisons and shifts.)

In the best-case scenario, where the data is already sorted in ascending order, we end up making just one comparison per passthrough, and not a single shift, since each value is already in its correct place.

Where data is randomly sorted, however, we'll have passthroughs in which we compare and shift all of the data, some of the data, or possibly none of data. If you'll look at our preceding walkthrough example, you'll notice that in Passthroughs #1 and #3, we compare and shift all the data we encounter. In Passthrough #4, we compare and shift some of it, and in Passthrough #2, we make just one comparison and shift no data at all.

While in the worst-case scenario, we compare and shift *all* the data, and in the best-case scenario, we shift *none* of the data (and just make one comparison per passthrough), for the average scenario, we can say that in the aggregate, we probably compare and shift about *half* of the data.

So if Insertion Sort takes N^2 steps for the worst-case scenario, we'd say that it takes about $N^2 / 2$ steps for the average scenario. (In terms of Big O, however, both scenarios are $O(N^2)$.)

Let's dive into some specific examples.

The array [1, 2, 3, 4] is already presorted, which is the best case. The worst case for the same data would be [4, 3, 2, 1], and an example of an average case might be [1, 3, 4, 2].

In the worst case, there are six comparisons and six shifts, for a total of twelve steps. In the average case, there are four comparisons and two shifts, for a total of six steps. In the best case, there are three comparisons and zero shifts.

We can now see that the performance of Insertion Sort *varies greatly* based on the scenario. In the worst-case scenario, Insertion Sort takes N^2 steps. In an average scenario, it takes $N^2 / 2$ steps. And in the best-case scenario, it takes about N steps.

This variance is because some passthroughs compare all the data to the left of the temp_value, while other passthroughs end early, due to encountering a value that is less than the temp_value.

You can see these three types of performance in this graph:

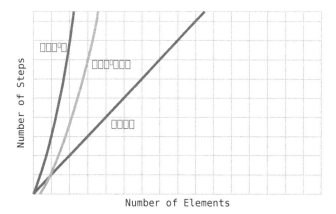

Compare this with Selection Sort. Selection Sort takes $N^2 / 2$ steps in *all* cases, from worst to average to best-case scenarios. This is because Selection Sort doesn't have any mechanism for ending a passthrough early at any point. Each passthrough compares every value to the right of the chosen index no matter what.

So which is better: Selection Sort or Insertion Sort? The answer is: well, it depends. In an average case—where an array is randomly sorted—they perform similarly. If you have reason to assume that you'll be dealing with data that is *mostly* sorted, Insertion Sort will be a better choice. If you have reason to assume that you'll be dealing with data that is mostly sorted in reverse order, Selection Sort will be faster. If you have no idea what the data will be like, that's essentially an average case, and both will be equal.

A Practical Example

Let's say that you are writing a JavaScript application, and somewhere in your code you find that you need to get the intersection between two arrays. The intersection is a list of all the values that occur in *both* of the arrays. For example, if you have the arrays [3, 1, 4, 2] and [4, 5, 3, 6], the intersection would be a third array, [3, 4], since both of those values are common to the two arrays.

JavaScript does not come with such a function built in, so we'll have to create our own. Here's one possible implementation:

```javascript
function intersection(first_array, second_array){
    var result = [];

    for (var i = 0; i < first_array.length; i++) {
        for (var j = 0; j < second_array.length; j++) {
            if (first_array[i] == second_array[j]) {
                result.push(first_array[i]);
            }
        }
    }
    return result;
}
```

Here, we are running a simple nested loop. In the outer loop, we iterate over each value in the first array. As we point to each value in the first array, we then run an inner loop that checks each value of the second array to see if it can find a match with the value being pointed to in the first array.

There are two types of steps taking place in this algorithm: comparisons and insertions. We compare every value of the two arrays against each other, and we insert matching values into the array result. The insertions are negligible beside the comparisons, since even in a scenario where the two arrays are identical, we'll only insert as many values as there are in one of the arrays. The primary steps to analyze, then, are the comparisons.

If the two arrays are of equal size, the number of comparisons performed are N^2. This is because for each element of the first array, we make a comparison of that element to each element of the second array. Thus, if we'd have two arrays each containing five elements, we'd end up making twenty-five comparisons. So this intersection algorithm has an efficiency of $O(N^2)$.

(If the arrays are different sizes—say N and M—we'd say that the efficiency of this function is $O(N * M)$. To simplify this example, however, let's assume that both arrays are of equal size.)

Is there any way we can improve this algorithm?

This is where it's important to consider scenarios beyond the worst case. In the current implementation of the intersection function, we make N^2 comparisons in every type of case, from where we have two arrays that do not contain a single common value, to a case of two arrays that are completely identical.

The truth is that in a case where the two arrays have no common value, we really have no choice but to check each and every value of both arrays to determine that there are no intersecting values.

But where the two arrays share common elements, we really shouldn't have to check every element of the first array against every element of the second array. Let's see why.

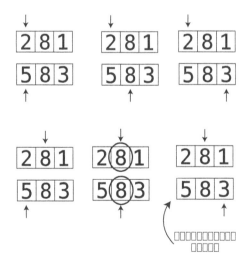

In this example, as soon as we find a common value (the 8), there's really no reason to complete the second loop. What are we checking for at this point? We've already determined that the second array contains the same element as the value we're pointing to in the first array. We're performing an unnecessary step.

To fix this, we can add a single word to our implementation:

```
function intersection(first_array, second_array){
    var result = [];

    for (var i = 0; i < first_array.length; i++) {
        for (var j = 0; j < second_array.length; j++) {
            if (first_array[i] == second_array[j]) {
                result.push(first_array[i]);
                break;
            }
        }
    }
    return result;
}
```

With the addition of the break, we can cut the inner loop short, and save steps (and therefore time).

Now, in a worst-case scenario, where the two elements do not contain a single shared value, we have no choice but to perform N^2 comparisons. In the best-case scenario, where the two arrays are identical, we only have to perform N comparisons. In an average case, where the two arrays are different but share *some* values, the performance will be somewhere in between N and N^2.

This is a significant optimization to our intersection function, since our first implementation would make N^2 comparisons in all cases.

Wrapping Up

Having the ability to discern between best-, average-, and worst-case scenarios is a key skill in choosing the best algorithm for your needs, as well as taking existing algorithms and optimizing them further to make them significantly faster. Remember, while it's good to be prepared for the worst case, average cases are what happen most of the time.

In the next chapter, we're going to learn about a new data structure that is similar to the array, but has nuances that can allow it to be more performant in certain circumstances. Just as you now have the ability to choose between two competing algorithms for a given use case, you'll also need the ability to choose between two competing data structures, as one may have better performance than the other.

CHAPTER 7

Blazing Fast Lookup with Hash Tables

Imagine that you're writing a program that allows customers to order food to go from a fast-food restaurant, and you're implementing a menu of foods with their respective prices. You could, of course, use an array:

```
menu = [ ["french fries", 0.75], ["hamburger", 2.5],
["hot dog", 1.5], ["soda", 0.6] ]
```

This array contains several subarrays, and each subarray contains two elements. The first element is a string representing the food on the menu, and the second element represents the price of that food.

As we learned in *Why Algorithms Matter*, if this array were unordered, searching for the price of a given food would take O(N) steps since the computer would have to perform a linear search. If it's an *ordered* array, the computer could do a binary search, which would take O(log N).

While O(log N) isn't bad, we can do better. In fact, we can do *much* better. By the end of this chapter, you'll learn how to use a special data structure called a *hash table*, which can be used to look up data in just O(1). By knowing how hash tables work under the hood and the right places to use them, you can leverage their tremendous lookup speeds in many situations.

Enter the Hash Table

Most programming languages include a data structure called a *hash table*, and it has an amazing superpower: fast reading. Note that hash tables are called by different names in various programming languages. Other names include hashes, maps, hash maps, dictionaries, and associative arrays.

Here's an example of the menu as implemented with a hash table using Ruby:

```
menu = { "french fries" => 0.75, "hamburger" => 2.5,
"hot dog" => 1.5, "soda" => 0.6 }
```

A hash table is a list of paired values. The first item is known as the *key*, and the second item is known as the *value*. In a hash table, the key and value have some significant association with one another. In this example, the string "french fries" is the key, and 0.75 is the value. They are paired together to indicate that french fries cost 75 cents.

In Ruby, you can look up a key's value using this syntax:

```
menu["french fries"]
```

This would return the value of 0.75.

Looking up a value in a hash table has an efficiency of O(1) on average, as it takes *just one step*. Let's see why.

Hashing with Hash Functions

Do you remember those secret codes you used as a kid to create and decipher messages?

For example, here's a simple way to map letters to numbers:

A = 1

B = 2

C = 3

D = 4

E = 5

and so on.

According to this code,

ACE converts to 135,

CAB converts to 312,

DAB converts to 412,

and

BAD converts to 214.

This process of taking characters and converting them to numbers is known as *hashing*. And the code that is used to convert those letters into particular numbers is called a *hash function*.

There are many other hash functions besides this one. Another example of a hash function is to take each letter's corresponding number and return the *sum* of all the numbers. If we did that, BAD would become the number 7 following a two-step process:

Step #1: First, BAD converts to 214.

Step #2: We then take each of these digits and get their sum:

2 + 1 + 4 = 7

Another example of a hash function is to return the *product* of all the letters' corresponding numbers. This would convert the word BAD into the number 8:

Step #1: First, BAD converts to 214.

Step #2: We then take the product of these digits:

2 * 1 * 4 = 8

In our examples for the remainder of this chapter, we're going to stick with this last version of the hash function. Real-world hash functions are more complex than this, but this "multiplication" hash function will keep our examples clear and simple.

The truth is that a hash function needs to meet only one criterion to be valid: a hash function must convert the same string to the *same number* every single time it's applied. If the hash function can return inconsistent results for a given string, it's not valid.

Examples of invalid hash functions include functions that use random numbers or the current time as part of their calculation. With these functions, BAD might convert to 12 one time, and 106 another time.

With our "multiplication" hash function, however, BAD will *always* convert to 8. That's because B is always 2, A is always 1, and D is always 4. And 2 * 1 * 4 is *always* 8. There's no way around this.

Note that with this hash function, DAB will *also* convert into 8 just as BAD will. This will actually cause some issues that we'll address later.

Armed with the concept of hash functions, we can now understand how a hash table actually works.

Building a Thesaurus for Fun and Profit, but Mainly Profit

On nights and weekends, you're single-handedly working on a stealth startup that will take over the world. It's...a thesaurus app. But this isn't any *old* thesaurus app—this is Quickasaurus. And you know that it will totally disrupt the billion-dollar thesaurus market. When a user looks up a word in Quickasaurus, it returns the word's most *popular* synonym, instead of *every* possible synonym as old-fashioned thesaurus apps do.

Since every word has an associated synonym, this is a great use case for a hash table. After all, a hash table is a list of paired items. So let's get started.

We can represent our thesaurus using a hash table:

thesaurus = {}

Under the hood, a hash table stores its data in a bunch of cells in a row, similar to an array. Each cell has a corresponding number. For example:

Let's add our first entry into the hash:

thesaurus["bad"] = "evil"

In code, our hash table now looks like this:

{"bad" => "evil"}

Let's explore how the hash table stores this data.

First, the computer applies the hash function to the key. Again, we'll be using the "multiplication" hash function described previously for demonstration purposes.

BAD = 2 * 1 * 4 = 8

Since our key ("bad") hashes into 8, the computer places the value ("evil") into cell 8:

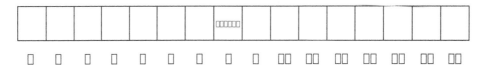

Now, let's add another key/value pair:

thesaurus["cab"] = "taxi"

Again, the computer hashes the key:

CAB = 3 * 1 * 2 = 6

Since the resulting value is 6, the computer stores the value ("taxi") inside cell 6.

Let's add one more key/value pair:

thesaurus["ace"] = "star"

ACE hashes into 15, since ACE = 1 * 3 * 5 = 15, so "star" gets placed into cell 15:

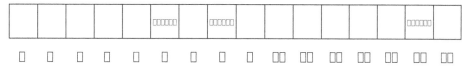

In code, our hash table currently looks like this:

{"bad" => "evil", "cab" => "taxi", "ace" => "star"}

Now that we've set up our hash table, let's see what happens when we look up values from it. Let's say we want to look up the value associated with the key of "bad". In our code, we'd say:

thesaurus["bad"]

The computer then executes two simple steps:

1. The computer hashes the key we're looking up: BAD = 2 * 1 * 4 = 8.
2. Since the result is 8, the computer looks inside cell 8 and returns the value that is stored there. In this case, that would be the string "evil".

It now becomes clear why looking up a value in a hash table is typically O(1): it's a process that takes a constant amount of time. The computer hashes the key we're looking up, gets the corresponding value, and jumps to the cell with that value.

We can now understand why a hash table would yield faster lookups for our restaurant menu than an array. With an array, were we to look up the price of a menu item, we would have to search through each cell until we found it. For an unordered array, this would take up to O(N), and for an ordered array, this would take up to O(log N). Using a hash table, however, we can now use the actual menu items as keys, allowing us to do a hash table lookup of O(1). And *that's* the beauty of a hash table.

Dealing with Collisions

Of course, hash tables are not without their complications.

Continuing our thesaurus example: what happens if we want to add the following entry into our thesaurus?

thesaurus["dab"] = "pat"

First, the computer would hash the key:

DAB = 4 * 1 * 2 = 8

And then it would try to add "pat" to our hash table's cell 8:

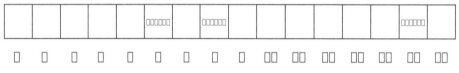

Uh oh. Cell 8 is already filled with "evil"—literally!

Trying to add data to a cell that is already filled is known as *collision*. Fortunately, there are ways around it.

One classic approach for handling collisions is known as *separate chaining*. When a collision occurs, instead of placing a *single* value in the cell, it places in it a reference to an array.

Let's look more carefully at a subsection of our hash table's underlying data storage:

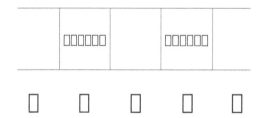

In our example, the computer wants to add "pat" to cell 8, but it already contains "evil". So it replaces the contents of cell 8 with an array as shown on page 83.

This array contains subarrays where the first value is the word, and the second word is its synonym.

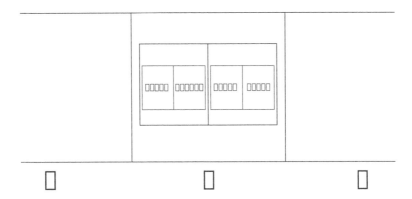

Let's walk through how a hash table lookup works in this case. If we look up:

thesaurus["dab"]

the computer takes the following steps:

1. It hashes the key. DAB = 4 * 1 * 2 = 8.

2. It looks up cell 8. The computer takes note that cell 8 contains an array of arrays rather than a single value.

3. It searches through the array linearly, looking at index 0 of each subarray until it finds the word we're looking up ("dab"). It then returns the value at index 1 of the correct subarray.

Let's walk through these steps visually.

We hash DAB into 8, so the computer inspects that cell:

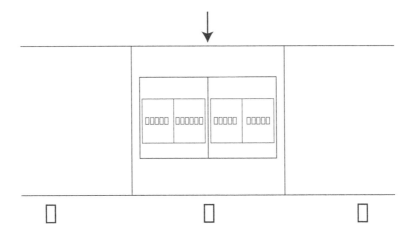

Since cell 8 contains an array, we begin a linear search through each cell, starting at the first one. It contains another array, and we inspect index 0:

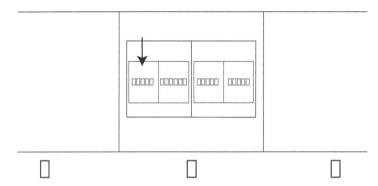

It does not contain the key we are looking for ("dab"), so we move on to the next cell:

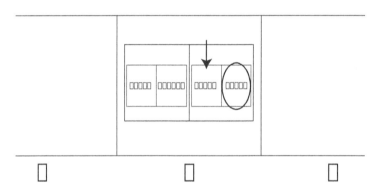

We found "dab", which would indicate that the value at index 1 of that subarray ("pat") is the value we're looking for.

In a scenario where the computer hits upon a cell that references an array, its search can take some extra steps, as it needs to conduct a linear search within an array of multiple values. If somehow all of our data ended up within a single cell of our hash table, our hash table would be no better than an array. So it actually turns out that the worst-case performance for a hash table lookup is O(N).

Because of this, it is critical that a hash table be designed in a way that it will have few collisions, and therefore typically perform lookups in O(1) time rather than O(N) time.

Let's see how hash tables are implemented in the real world to avoid frequent collisions.

The Great Balancing Act

Ultimately, a hash table's efficiency depends on three factors:

- How much data we're storing in the hash table
- How many cells are available in the hash table
- Which hash function we're using

It makes sense why the first two factors are important. If you have a lot of data to store in only a few cells, there will be many collisions and the hash table will lose its efficiency. Let's explore, however, why the hash function itself is important for efficiency.

Let's say that we're using a hash function that always produces a value that falls in the range between 1 and 9 inclusive. An example of this is a hash function that converts letters into their corresponding numbers, and keeps adding the resulting digits together until it ends up with a single digit.

For example:

PUT = 16 + 21 + 20 = 57

Since 57 contains more than one digit, the hash function breaks up the 57 into 5 + 7:

5 + 7 = 12

12 also contains more than one digit, so it breaks up the 12 into 1 + 2:

1 + 2 = 3

In the end, PUT hashes into 3. This hash function by its very nature will *always* return a number 1 through 9.

Let's return to our example hash table:

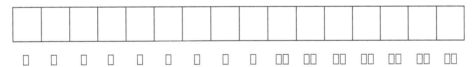

With this hash function, the computer would never even use cells 10 through 16 even though they exist. All data would be stuffed into cells 1 through 9.

A good hash function, therefore, is one that distributes its data across all available cells.

If we need a hash table to store just five values, how big should our hash table be, and what type of hash function should we use?

If a hash table had only five cells, we'd need a hash function that converts keys into numbers 1 through 5. Even if we only planned on storing five pieces of data, there's a good chance that there will be a collision or two, since two keys may be easily hashed to the same value.

However, if our hash table were to have *one hundred* cells, and our hash function converts strings into numbers 1 through 100, when storing just five values it would be much less likely to have any collisions since there are one hundred possible cells that each of those strings might end up in.

Although a hash table with one hundred cells is great for avoiding collisions, we'd be using up one hundred cells to store just five pieces of data, and that's a poor use of memory.

This is the balancing act that a hash table must perform. A good hash table *strikes a balance of avoiding collisions while not consuming lots of memory.*

To accomplish this, computer scientists have developed the following rule of thumb: for every seven data elements stored in a hash table, it should have ten cells.

So, if you're planning on storing fourteen elements, you'd want to have twenty available cells, and so on and so forth.

This ratio of data to cells is called the *load factor*. Using this terminology, we'd say that the ideal load factor is 0.7 (7 elements / 10 cells).

If you initially stored seven pieces of data in a hash, the computer might allocate a hash with ten cells. When you begin to add more data, though, the computer will expand the hash by adding more cells and changing the hash function so that the new data will be distributed evenly across the new cells.

Luckily, most of the internals of a hash table are managed by the computer language you're using. It decides how big the hash table needs to be, what hash function to use, and when it's time to expand the hash table. But now that you understand *how* hashes work, you can use them to replace arrays in many cases to optimize your code's performance and take advantage of superior lookups that have an efficiency of O(1).

Practical Examples

Hash tables have many practical use cases, but here we're going to focus on using them to increase algorithm speed.

In *Why Data Structures Matter*, we learned about array-based sets—arrays that ensure that no duplicate values exist. There we found that every time a new value is inserted into the set, we need to run a linear search (if the set is unordered) to make sure that the value we're inserting isn't already in the set.

If we're dealing with a large set in which we're making lots of insertions, this can get unwieldy very quickly, since every insertion runs at O(N), which isn't very fast.

In many cases, we can use a hash table to serve as a set.

When we use an array as a set, we simply place each piece of data in a cell within the array. When we use a hash table as a set, however, each piece of data is a key within the hash table, and the value can be anything, such as a 1 or a Boolean value of true.

Let's say we set up our hash table set in JavaScript as follows:

```
var set = {};
```

Let's add a few values to the set:

```
set["apple"] = 1;
set["banana"] = 1;
set["cucumber"] = 1;
```

Every time we want to insert a new value, instead of a O(N) linear search, we can simply add it in O(1) time. This is true even if we add a key that already exists:

```
set["banana"] = 1;
```

When we attempt to add another "banana" to the set, we don't need to check whether "banana" already exists, since even if it does, we're simply overwriting the value for that key with the number 1.

The truth is that hash tables are perfect for any situation where we want to keep track of which values exist within a dataset. In *Speeding Up Your Code with Big O*, we discussed writing a JavaScript function that would tell us whether any duplicate numbers occur within an array. The first solution we provided was:

```
function hasDuplicateValue(array) {
    for(var i = 0; i < array.length; i++) {
        for(var j = 0; j < array.length; j++) {
            if(i !== j && array[i] == array[j]) {
                return true;
            }
        }
    }
    return false;
}
```

We noted there that this code, which contains nested loops, runs at O(N²).

The second solution we suggested there runs at O(N), but only works if the array exclusively contains positive integers. What if the array contains something else, like strings?

With a hash table (which is called an *object* in JavaScript), we can achieve a similar solution that would effectively handle strings:

```
function hasDuplicateValue(array) {
    var existingValues = {};
    for(var i = 0; i < array.length; i++) {
        if(existingValues[array[i]] === undefined) {
            existingValues[array[i]] = 1;
        } else {
            return true;
        }
    }
    return false;
}
```

This approach also runs at O(N). Here existingValues is a hash table rather than an array, so its keys can include strings as well as integers.

Let's say we are building an electronic voting machine, in which voters can choose from a list of candidates or write in another candidate. If the only time we counted the final tally of votes was at the end of the election, we could store the votes as a simple array, and insert each vote at the end as it comes in:

```
var votes = [];
function addVote(candidate) {
  votes.push(candidate);
}
```

We'd end up with a very long array that would look something like

```
["Thomas Jefferson", "John Adams", "John Adams", "Thomas Jefferson",
 "John Adams", ...]
```

With this approach, each insertion would only take O(1).

But what about the final tally? Well, we could count the votes by running a loop and keeping track of the results in a hash table:

```
function countVotes(votes) {
  var tally = {};
  for(var i = 0; i < votes.length; i++) {
    if(tally[votes[i]]) {
      tally[votes[i]]++;
```

```
    } else {
      tally[votes[i]] = 1;
    }
  }

  return tally;
}
```

However, since we'd have to count each vote at the end of the day, the countVotes function would take O(N). This might take too long!

Instead, we may consider using the hash table to store the data in the first place:

```
var votes = {};

function addVote(candidate) {
  if(votes[candidate]) {
    votes[candidate]++;
  } else {
    votes[candidate] = 1;
  }
}

function countVotes() {
  return votes;
}
```

With this technique, not only are insertions O(1), but our tally is as well, since it's already kept as the voting takes place.

Wrapping Up

Hash tables are indispensable when it comes to building efficient software. With their O(1) reads and insertions, it's a difficult data structure to beat.

Until now, our analysis of various data structures revolved around their efficiency and speed. But did you know that some data structures provide advantages other than speed? In the next lesson, we're going to explore two data structures that can help improve code elegance and maintainability.

CHAPTER 8

Crafting Elegant Code with Stacks and Queues

Until now, our discussion around data structures has focused primarily on how they affect the *performance* of various operations. However, having a variety of data structures in your programming arsenal allows you to create code that is simpler and easier to read.

In this chapter, you're going to discover two new data structures: stacks and queues. The truth is that these two structures are not entirely new. They're simply arrays with restrictions. Yet, these restrictions are exactly what make them so elegant.

More specifically, stacks and queues are elegant tools for handling temporary data. From operating system architecture to printing jobs to traversing data, stacks and queues serve as temporary containers that can be used to form beautiful algorithms.

Think of temporary data like the food orders in a diner. What each customer orders is important until the meal is made and delivered; then you throw the order slip away. You don't need to keep that information around. Temporary data is information that doesn't have any meaning after it's processed, so you don't need to keep it around. However, the order in which you process the data can be important—in the diner, you should ideally make each meal in the order in which it was requested. Stacks and queues allow you to handle data in order, and then get rid of it once you don't need it anymore.

Stacks

A *stack* stores data in the same way that arrays do—it's simply a list of elements. The one catch is that stacks have the following three constraints:

- Data can only be inserted at the end of a stack.
- Data can only be read from the end of a stack.
- Data can only be removed from the end of a stack.

You can think of a stack as an actual stack of dishes: you can't look at the face of any dish other than the one at the top. Similarly, you can't add any dish except to the top of the stack, nor can you remove any dish besides the one at the top. (At least, you shouldn't.) In fact, most computer science literature refers to the end of the stack as its *top*, and the beginning of the stack as its *bottom*.

While these restrictions seem—well—restrictive, we'll see shortly how they are to our benefit.

To see a stack in action, let's start with an empty stack.

Inserting a new value into a stack is also called *pushing onto the stack*. Think of it as adding a dish onto the top of the dish stack.

Let's push a 5 onto the stack:

Again, there's nothing fancy going on here. We're just inserting data into the end of an array.

Now, let's push a 3 onto the stack:

Next, let's push a 0 onto the stack:

Note that we're always adding data to the top (that is, the end) of the stack. If we wanted to insert the 0 into the bottom or middle of the stack, we couldn't, because that is the nature of a stack: data can only be added to the top.

Removing elements from the top of the stack is called *popping from the stack*. Because of a stack's restrictions, we can only pop data from the top.

Let's pop some elements from our example stack:

First we pop the 0:

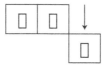

Our stack now contains only two elements, the 5 and the 3.

Next, we pop the 3:

Our stack now only contains the 5:

A handy acronym used to describe stack operations is *LIFO*, which stands for "Last In, First Out." All this means is that the *last* item *pushed* onto a stack is always the *first* item *popped* from it. It's sort of like students who slack off—they're always the last to arrive to class, but the first to leave.

Stacks in Action

Although a stack is not typically used to store data on a long-term basis, it can be a great tool to handle temporary data as part of various algorithms. Here's an example:

Let's create the beginnings of a JavaScript linter—that is, a program that inspects a programmer's JavaScript code and ensures that each line is syntactically correct. It can be very complicated to create a linter, as there are many different aspects of syntax to inspect. We'll focus on just one specific aspect of the linter—opening and closing braces. This includes parentheses, square brackets, and curly braces—all common causes of frustrating syntax errors.

To solve this problem, let's first analyze what type of syntax is incorrect when it comes to braces. If we break it down, we'll find that there are three situations of erroneous syntax:

The first is when there is an opening brace that doesn't have a corresponding closing brace, such as this:

(var x = 2;

We'll call this Syntax Error Type #1.

The second is when there is a closing brace that was never preceded by a corresponding opening brace:

var x = 2;)

We'll call that Syntax Error Type #2.

The third, which we'll refer to as Syntax Error Type #3, is when a closing brace is not the same *type* of brace as the immediately preceding opening brace, such as:

(var x = [1, 2, 3)];

In the preceding example, there is a matching set of parentheses, and a matching pair of square brackets, but the closing parenthesis is in the wrong place, as it doesn't match the immediately preceding opening brace, which is a square bracket.

How can we implement an algorithm that inspects a line of code and ensures that there are no syntax errors in this regard? This is where a stack allows us to implement a beautiful linting algorithm, which works as follows:

We prepare an empty stack, and then we read each character from left to right, following these rules:

1. If we find any character that isn't a type of brace (parenthesis, square bracket, or curly brace), we ignore it and move on.

2. If we find an *opening* brace, we push it onto the stack. Having it on the stack means we're waiting to close that particular brace.

3. If we find a *closing* brace, we inspect the top element in the stack. We then analyze:

    - If there are no elements in the stack, that means we've found a closing brace without a corresponding opening brace beforehand. This is Syntax Error Type #2.

    - If there *is* data in the stack, but the closing brace is *not* a corresponding match for the top element of the stack, that means we've encountered Syntax Error Type #3.

- If the closing character *is* a corresponding match for the element at the top of the stack, that means we've successfully closed that opening brace. We pop the top element from the stack, since we no longer need to keep track of it.

4. If we make it to the end of the line and there's still something left on the stack, that means there's an opening brace without a corresponding closing brace, which is Syntax Error Type #1.

Let's see this in action using the following example:

$$(var\ x\ =\ \{y:\ [1,\ 2,\ 3]\})$$

After we prepare an empty stack, we begin reading each character from left to right:

Step #1: We begin with the first character, which happens to be an opening parenthesis:

$$\downarrow$$
$$(var\ x\ =\ \{y:\ [1,\ 2,\ 3]\})$$

Step #2: Since it's a type of opening brace, we push it onto the stack:

We then ignore all the characters var x = , since they aren't brace characters.

Step #3: We encounter our next opening brace:

$$\downarrow$$
$$(var\ x\ =\ \{y:\ [1,\ 2,\ 3]\})$$

Step #4: We push it onto the stack:

We then ignore the y:

Step #5: We encounter the opening square bracket:

$$\downarrow$$
$$(var\ x\ =\ \{y:\ [1,\ 2,\ 3]\})$$

Step #6: We add that to the stack as well:

We then ignore the 1, 2, 3.

Step #7: We encounter our first closing brace—a closing square bracket:

(var x = {y: [1, 2, 3]})

Step #8: We inspect the element at the top of the stack, which happens to be an *opening* square bracket. Since our closing square bracket is a corresponding match to this final element of the stack, we pop the opening square bracket from the stack:

Step #9: We move on, encountering a closing curly brace:

(var x = {y: [1, 2, 3]})

Step #10: We read the last element in the stack. It's an opening curly brace, so we've found a match. We pop the opening curly brace from our stack:

Step #11: We encounter a closing parenthesis:

(var x = {y: [1, 2, 3]})

Step #12: We read the last element in the stack. It's a corresponding match, so we pop it from the stack, leaving us with an empty stack.

Since we've made it through the entire line of code, and our stack is empty, our linter can conclude that there are no syntactical errors on this line (that relate to opening and closing braces).

Here's a Ruby implementation of the preceding algorithm. Note that Ruby has push and pop methods built into arrays, and they are essentially shortcuts for adding an element to the end of the array and removing an element from the end of an array, respectively. By only using those two methods to add and remove data from the array, we effectively use the array as a stack.

```
class Linter
  attr_reader :error
  def initialize
```

```ruby
      # We use a simple array to serve as our stack:
      @stack = []
    end

    def lint(text)
      # We start a loop which reads each character in our text:
      text.each_char.with_index do |char, index|

        if opening_brace?(char)

          # If the character is an opening brace, we push it onto the stack:
          @stack.push(char)
        elsif closing_brace?(char)

          if closes_most_recent_opening_brace?(char)

            # If the character closes the most recent opening brace,
            # we pop the opening brace from our stack:
            @stack.pop

          else # if the character does NOT close the most recent opening brace

            @error = "Incorrect closing brace: #{char} at index #{index}"
            return
          end
        end
      end

      if @stack.any?

        # If we get to the end of line, and the stack isn't empty, that means
        # we have an opening brace without a corresponding closing brace:
        @error = "#{@stack.last} does not have a closing brace"
      end
    end

    private

    def opening_brace?(char)
      ["(", "[", "{"].include?(char)
    end

    def closing_brace?(char)
      [")", "]", "}"].include?(char)
    end

    def opening_brace_of(char)
      {")" => "(", "]" => "[", "}" => "{"}[char]
    end

    def most_recent_opening_brace
      @stack.last
    end

    def closes_most_recent_opening_brace?(char)
      opening_brace_of(char) == most_recent_opening_brace
    end
end
```

If we use this class as follows:

```
linter = Linter.new
linter.lint("( var x = { y: [1, 2, 3] } )")
puts linter.error
```

We will not receive any errors since this line of JavaScript is correct. However, if we accidentally swap the last two characters:

```
linter = Linter.new
linter.lint("( var x = { y: [1, 2, 3] ) }")
puts linter.error
```

We will then get the message:

```
Incorrect closing brace: ) at index 25
```

If we leave off the last closing parenthesis altogether:

```
linter = Linter.new
linter.lint("( var x = { y: [1, 2, 3] }")
puts linter.error
```

We'll get this error message:

```
( does not have a closing brace
```

In this example, we use a stack to elegantly keep track of opening braces that have not yet been closed. In the next chapter, we'll use a stack to similarly keep track of functions of code that need to be invoked, which is the key to a critical concept known as recursion.

Stacks are ideal for processing *any* data that should be handled in reverse order to how it was received (LIFO). The "undo" function in a word processor, or function calls in a networked application, are examples of when you'd want to use a stack.

Queues

A *queue* also deals with temporary data elegantly, and is like a stack in that it is an array with restrictions. The difference lies in what order we want to process our data, and this depends on the particular application we are working on.

You can think of a queue as a line of people at the movie theater. The first one on the line is the first one to leave the line and enter the theater. With queues, the first item added to the queue is the first item to be removed. That's why computer scientists use the acronym "FIFO" when it comes to queues: First In, First Out.

Like stacks, queues are arrays with three restrictions (it's just a different set of restrictions):

- Data can only be inserted at the *end* of a queue. (This is identical behavior as the stack.)

- Data can only be read from the *front* of a queue. (This is the opposite of behavior of the stack.)

- Data can only be removed from the *front* of a queue. (This, too, is the opposite behavior of the stack.)

Let's see a queue in action, beginning with an empty queue.

First, we insert a 5: (While inserting into a stack is called *pushing*, inserting into a queue doesn't have a standardized name. Various references call it putting, adding, or enqueuing.)

Next, we insert a 9:

Next, we insert a 100:

As of now, the queue has functioned just like a stack. However, removing data happens in the reverse, as we remove data from the *front* of the queue.

If we want to remove data, we must start with the 5, since it's at the front of the queue:

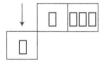

Next, we remove the 9:

Our queue now only contains one element, the 100.

Queues in Action

Queues are common in many applications, ranging from printing jobs to background workers in web applications.

Let's say you were programming a simple Ruby interface for a printer that can accept printing jobs from various computers across a network. By utilizing the Ruby array's `push` method, which adds data to the end of the array, and the `shift` method, which removes data from the beginning of the array, you may create a class like this:

```ruby
class PrintManager
  def initialize
    @queue = []
  end

  def queue_print_job(document)
    @queue.push(document)
  end

  def run
    while @queue.any?
      # the Ruby shift method removes and returns the
      # first element of an array:
      print(@queue.shift)
    end
  end

  private

  def print(document)
    # Code to run the actual printer goes here.
    # For demo purposes, we'll print to the terminal:
    puts document
  end
end
```

We can then utilize this class as follows:

```ruby
print_manager = PrintManager.new
print_manager.queue_print_job("First Document")
print_manager.queue_print_job("Second Document")
print_manager.queue_print_job("Third Document")
print_manager.run
```

The printer will then print the three documents in the same order in which they were received:

```
First Document
Second Document
Third Document
```

While this example is simplified and abstracts away some of the nitty-gritty details that a real live printing system may have to deal with, the fundamental use of a queue for such an application is very real and serves as the foundation for building such a system.

Queues are also the perfect tool for handling asynchronous requests—they ensure that the requests are processed in the order in which they were received. They are also commonly used to model real-world scenarios where events need to occur in a certain order, such as airplanes waiting for takeoff and patients waiting for their doctor.

Wrapping Up

As you've seen, stacks and queues are programmers' tools for elegantly handling all sorts of practical algorithms.

Now that we've learned about stacks and queues, we've unlocked a new achievement: we can learn about recursion, which depends upon a stack. Recursion also serves as the foundation for many of the more advanced and super efficient algorithms that we will cover in the rest of this book.

CHAPTER 9

Recursively Recurse with Recursion

You'll need to understand the key concept of *recursion* in order to understand most of the remaining algorithms in this book. Recursion allows for solving tricky problems in surprisingly simple ways, often allowing us to write a fraction of the code that we might otherwise write.

But first, a pop quiz!

What happens when the blah() function defined below is invoked?

```
function blah() {
  blah();
}
```

As you probably guessed, it will call itself infinitely, since blah() calls itself, which in turn calls itself, and so on.

Recursion is the official name for when a function calls itself. While infinite function calls are generally useless—and even dangerous—recursion is a powerful tool that can be harnessed. And when we harness the power of recursion, we can solve particularly tricky problems, as you'll now see.

Recurse Instead of Loop

Let's say you work at NASA and need to program a countdown function for launching spacecraft. The particular function that you're asked to write should accept a number—such as 10—and display the numbers from 10 down to 0. Take a moment and implement this function in the language of your choice. When you're done, read on.

Odds are that you wrote a simple loop, along the lines of this JavaScript implementation:

```
function countdown(number) {
  for(var i = number; i >= 0; i--) {
```

```
    console.log(i);
  }
}
countdown(10);
```

There's nothing wrong with this implementation, but it may have never occurred to you that you don't *have* to use a loop.

How?

Let's try recursion instead. Here's a first attempt at using recursion to implement our countdown function:

```
function countdown(number) {
  console.log(number);
  countdown(number - 1);
}
countdown(10);
```

Let's walk through this code step by step.

Step #1: We call countdown(10), so the argument number holds a 10.

Step #2: We print number (which contains the value 10) to the console.

Step #3: Before the countdown function is complete, it calls countdown(9) (since number - 1 is 9).

Step #4: countdown(9) begins running. In it, we print number (which is currently 9) to the console.

Step #5: Before countdown(9) is complete, it calls countdown(8).

Step #6: countdown(8) begins running. We print 8 to the console.

Before we continue stepping through the code, note how we're using recursion to achieve our goal. We're not using any loop constructs, but by simply having the countdown function call itself, we are able to count down from 10 and print each number to the console.

In almost any case in which you can use a loop, you can also use recursion. Now, just because you *can* use recursion doesn't mean that you *should* use recursion. Recursion is a tool that allows for writing elegant code. In the preceding example, the recursive approach is not necessarily any more beautiful or efficient than using a classic for loop. However, we will soon encounter examples in which recursion shines. In the meantime, let's continue exploring how recursion works.

The Base Case

Let's continue our walkthrough of the countdown function. We'll skip a few steps for brevity...

Step #21: We call countdown(0).

Step #22: We print number (i.e., 0) to the console.

Step #23: We call countdown(-1).

Step #24: We print number (i.e., -1) to the console.

Uh oh. As you can see, our solution is not perfect, as we'll end up stuck with infinitely printing negative numbers.

To perfect our solution, we need a way to end our countdown at 0 and prevent the recursion from continuing on forever.

We can solve this problem by adding a conditional statement that ensures that if number is currently 0, we don't call countdown() again:

```
function countdown(number) {
  console.log(number);
  if(number === 0) {
    return;
  } else {
    countdown(number - 1);
  }
}
countdown(10);
```

Now, when number is 0, our code will not call the countdown() function again, but instead just return, thereby preventing another call of countdown().

In Recursion Land (a real place), this case in which the method will *not* recurse is known as the *base case*. So in our countdown() function, 0 is the base case.

Reading Recursive Code

It takes time and practice to get used to recursion, and you will ultimately learn *two* sets of skills: *reading* recursive code, and *writing* recursive code. Reading recursive code is somewhat easier, so let's first get some practice with that.

Let's look at another example: calculating factorials.

A *factorial* is best illustrated with an example:

The factorial of 3 is:

3 * 2 * 1 = 6

The factorial of 5 is:

5 * 4 * 3 * 2 * 1 = 120

And so on and so forth. Here's a recursive implementation that returns a number's factorial using Ruby:

```ruby
def factorial(number)
  if number == 1
    return 1
  else
    return number * factorial(number - 1)
  end
end
```

This code can look somewhat confusing at first glance, but here's the process you need to know for reading recursive code:

1. Identify what the base case is.
2. Walk through the function assuming it's dealing with the base case.
3. Then, walk through the function assuming it's dealing with the case immediately *before* the base case.
4. Progress through your analysis by moving up the cases one at a time.

Let's apply this process to the preceding code. If we analyze the code, we'll quickly notice that there are two paths:

```ruby
if number == 1
  return 1
else
  return number * factorial(number - 1)
end
```

We can see that the recursion happens here, as factorial calls itself:

```ruby
else
  return number * factorial(number - 1)
end
```

So it must be that the following code refers to the base case, since this is the case in which the function does *not* call itself:

```ruby
if number == 1
  return 1
```

We can conclude, then, that number being 1 is the base case.

Next, let's walk through the factorial method assuming it's dealing with the base case, which would be factorial(1). The relevant code from our method is:

```
if number == 1
  return 1
```

Well, that's pretty simple—it's the base case, so no recursion actually happens. If we call factorial(1), the method simply returns 1. Okay, so grab a napkin and write this fact down:

factorial (1) returns 1

Next, let's move up to the next case, which would be factorial(2). The relevant line of code from our method is:

```
else
  return number * factorial(number - 1)
end
```

So, calling factorial(2) will return 2 * factorial(1). To calculate 2 * factorial(1), we need to know what factorial(1) returns. If you check your napkin, you'll see that it returns 1. So 2 * factorial(1) will return 2 * 1, which just happens to be 2.

Add this fact to your napkin:

factorial (2) returns 2

factorial (1) returns 1

Now, what happens if we call factorial(3)? Again, the relevant line of code is:

```
else
  return number * factorial(number - 1)
end
```

So that would translate into return 3 * factorial(2). What does factorial(2) return? You don't have to figure that out all over again, since it's on your napkin! It returns 2. So factorial(3) will return 6 (3 * 2 = 6). Go ahead and add this wonderful factoid to your napkin:

factorial (3) returns 6

factorial (2) returns 2

factorial (1) returns 1

Take a moment and figure out for yourself what factorial(4) will return.

As you can see, starting the analysis from the base case and building up is a great way to reason about recursive code.

In fact, following this process is not just a good way for humans to conceptualize recursion. Your computer uses a similar approach. Let's see how.

Recursion in the Eyes of the Computer

If you think about our factorial method, you'll realize when we call factorial(3), the following happens:

The computer calls factorial(3), and before the method is complete, it calls factorial(2), and before factorial(2) is complete, it calls factorial(1). Technically, while the computer runs factorial(1), it's still in the middle of factorial(2), which in turn is running within factorial(3).

The computer uses a stack to keep track of which functions it's in the middle of calling. This stack is known, appropriately enough, as the *call stack*.

Let's watch the call stack in action using the factorial example.

The computer begins by calling factorial(3). Before the method completes executing, however, factorial(2) gets called. In order to track that the computer is still in the middle of factorial(3), the computer pushes that info onto a call stack:

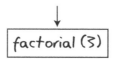

The computer then proceeds to execute factorial(2). Now, factorial(2), in turn, calls factorial(1). Before the computer dives into factorial(1), though, the computer needs to remember that it's still in the middle of factorial(2), so it pushes that onto the call stack as well:

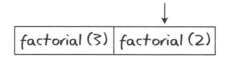

The computer then executes factorial(1). Since 1 is the base case, factorial(1) completes without the factorial method again.

Even after the computer completes factorial(1), it knows that it's not finished with everything it needs to do since there's data in the call stack, which indicates that it is still in the middle of running other methods that it needs to complete. As you will recall, stacks are restricted in that we can only inspect the top (in other words, the final) element. So the next thing the computer does is peek at the top element of the call stack, which currently is factorial(2).

Since factorial(2) is the last item in the call stack, that means that factorial(2) is the most recently called method and therefore the immediate method that needs to be completed.

The computer then pops factorial(2) from the call stack

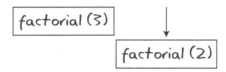

and completes running factorial(2).

Then the computer looks at the call stack to see which method it needs to complete next, and since the call stack is currently

$$\boxed{factorial\ (3)}$$

it pops factorial(3) from the stack and completes it.

At this point, the stack is empty, so the computer knows it's done executing all of the methods, and the recursion is complete.

Looking back at this example from a bird's-eye view, you'll see that the order in which the computer calculates the factorial of 3 is as follows:

1. factorial(3) is called first.
2. factorial(2) is called second.
3. factorial(1) is called third.
4. factorial(1) is *completed* first.
5. factorial(2) is completed based on the result of factorial(1).
6. Finally, factorial(3) is completed based on the result of factorial(2).

Interestingly, in the case of infinite recursion (such as the very first example in our chapter), the program keeps on pushing the same method over and over again onto the call stack, until there's no more room in the computer's memory—and this causes an error known as *stack overflow*.

Recursion in Action

While previous examples of the NASA countdown and calculating factorials can be solved with recursion, they can also be easily solved with classical loops. While recursion is interesting, it doesn't really provide an advantage when solving these problems.

However, recursion is a natural fit in any situation where you find yourself having to repeat an algorithm within the same algorithm. In these cases, recursion can make for more readable code, as you're about to see.

Take the example of traversing through a filesystem. Let's say that you have a script that does something with every file inside of a directory. However, you don't want the script to only deal with the files inside the *one* directory—you want it to act on all the files within the *subdirectories* of the directory, and the subdirectories of the subdirectories, and so on.

Let's create a simple Ruby script that prints out the names of all subdirectories within a given directory.

```ruby
def find_directories(directory)
  Dir.foreach(directory) do |filename|
    if File.directory?("#{directory}/#{filename}") &&
    filename != "." && filename != ".."
      puts "#{directory}/#{filename}"
    end
  end
end

# Call the find_directories method on the current directory:
find_directories(".")
```

In this script, we look through each file within the given directory. If the file is itself a subdirectory (and isn't a single or double period, which represent the current and previous directories, respectively), we print the subdirectory name.

While this works well, it only prints the names of the subdirectories *immediately* within the current directory. It does not print the names of the subdirectories *within* those subdirectories.

Let's update our script so that it can search one level deeper:

```ruby
def find_directories(directory)
  # Loop through outer directory:
  Dir.foreach(directory) do |filename|
    if File.directory?("#{directory}/#{filename}") &&
    filename != "." && filename != ".."
      puts "#{directory}/#{filename}"
```

```ruby
      # Loop through inner subdirectory:
      Dir.foreach("#{directory}/#{filename}") do |inner_filename|
        if File.directory?("#{directory}/#{filename}/#{inner_filename}") &&
          inner_filename != "." && inner_filename != ".."
            puts "#{directory}/#{filename}/#{inner_filename}"
        end
      end
    end
  end
end

# Call the find_directories method on the current directory:
find_directories(".")
```

Now, every time our script discovers a directory, it then conducts an identical loop through the subdirectories of *that* directory and prints out the names of the subdirectories. But this script also has its limitations, because it's only searching two levels deep. What if we wanted to search three, four, or five levels deep? What if we wanted to search as deep as our subdirectories go? That would seem to be impossible.

And *this* is the beauty of recursion. With recursion, we can write a script that goes arbitrarily deep—and is also simple!

```ruby
def find_directories(directory)
  Dir.foreach(directory) do |filename|
    if File.directory?("#{directory}/#{filename}") &&
      filename != "." && filename != ".."
        puts "#{directory}/#{filename}"
        find_directories("#{directory}/#{filename}")
    end
  end
end

# Call the find_directories method on the current directory:
find_directories(".")
```

As this script encounters files that are themselves subdirectories, it calls the find_directories method upon that very subdirectory. The script can therefore dig as deep as it needs to, leaving no subdirectory unturned.

To visually see how this algorithm applies to an example filesystem, examine the diagram on page 112, which specifies the order in which the script traverses the subdirectories.

Note that recursion in a vacuum does not necessarily speed up an algorithm's efficiency in terms of Big O. However, we will see in the following chapter that recursion can be a core component of algorithms that *does* affect their speed.

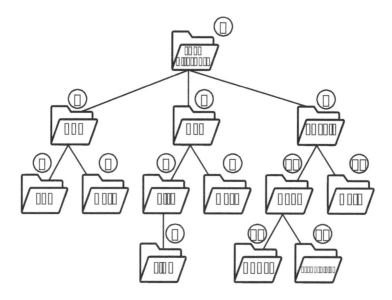

Wrapping Up

As you've seen in the filesystem example, recursion is often a great choice for an algorithm in which the algorithm itself doesn't know on the outset how many levels deep into something it needs to dig.

Now that you understand recursion, you've also unlocked a superpower. You're about to encounter some really efficient—yet advanced—algorithms, and many of them rely on the principles of recursion.

CHAPTER 10

Recursive Algorithms for Speed

We've seen that understanding recursion unlocks all sorts of new algorithms, such as searching through a filesystem. In this chapter, we're going to learn that recursion is also the key to algorithms that can make our code run much, much faster.

In previous chapters, we've encountered a number of sorting algorithms, including Bubble Sort, Selection Sort, and Insertion Sort. In real life, however, none of these methods are actually used to sort arrays. Most computer languages have built-in sorting functions for arrays that save us the time and effort from implementing our own. And in many of these languages, the sorting algorithm that is employed under the hood is *Quicksort*.

The reason we're going to dig into Quicksort even though it's built in to many computer languages is because by studying how it works, we can learn how to use recursion to greatly speed up an algorithm, and we can do the same for other practical algorithms of the real world.

Quicksort is an extremely fast sorting algorithm that is particularly efficient for average scenarios. While in worst-case scenarios (that is, inversely sorted arrays), it performs similarly to Insertion Sort and Selection Sort, it is much faster for average scenarios—which are what occur most of the time.

Quicksort relies on a concept called *partitioning*, so we'll jump into that first.

Partitioning

To *partition* an array is to take a random value from the array—which is then called the *pivot*—and make sure that every number that is less than the pivot ends up to the left of the pivot, and that every number that is greater than the pivot ends up to the right of the pivot. We accomplish partitioning through a simple algorithm that will be described in the following example.

Let's say we have the following array:

For consistency's sake, we'll always select the right-most value to be our pivot (although we can technically choose other values). In this case, the number 3 is our pivot. We indicate this by circling it:

We then assign "pointers"—one to the left-most value of the array, and one to the right-most value of the array, excluding the pivot itself:

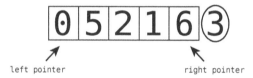

We're now ready to begin the actual partition, which follows these steps:

1. The left pointer continuously moves one cell to the right until it reaches a value that is greater than or equal to the pivot, and then stops.

2. Then, the right pointer continuously moves one cell to the left until it reaches a value that is less than or equal to the pivot, and then stops.

3. We swap the values that the left and right pointers are pointing to.

4. We continue this process until the pointers are pointing to the very same value or the left pointer has moved to the right of the right pointer.

5. Finally, we swap the pivot with the value that the left pointer is currently pointing to.

When we're done with a partition, we are now assured that all values to the left of the pivot are less than the pivot, and all values to the right of the pivot are greater than it. And that means that the pivot itself is now in its correct place within the array, although the other values are not yet necessarily completely sorted.

Let's apply this to our example:

Step #1: Compare the left pointer (now pointing to 0) to our pivot (the value 3):

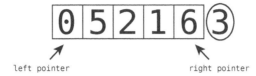

Since 0 is less than the pivot, the left pointer moves on in the next step.

Step #2: The left pointer moves on:

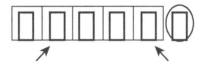

We compare the left pointer (the 5) to our pivot. Is the 5 lower than the pivot? It's not, so the left pointer stops, and we begin to activate the right pointer in our next step.

Step #3: Compare the right pointer (6) to our pivot. Is the value greater than the pivot? It is, so our pointer will move on in the next step.

Step #4: The right pointer moves on:

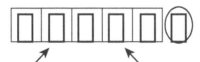

We compare the right pointer (1) to our pivot. Is the value greater than the pivot? It's not, so our right pointer stops.

Step #5: Since both pointers have stopped, we swap the values of the two pointers:

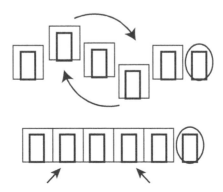

We then activate our left pointer again in the next step.

Step #6: The left pointer moves on:

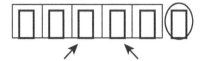

We compare the left pointer (2) to our pivot. Is the value less than the pivot? It is, so the left pointer moves on.

Step #7: The left pointer moves on to the next cell. Note that at this point, both the left and right pointers are pointing to the same value:

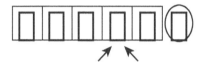

We compare the left pointer to our pivot. Since our left pointer is pointing to a value that is greater than our pivot, it stops. At this point, since our left pointer has reached our right pointer, we are done with moving pointers.

Step #8: For our final step of the partition, we swap the value that the left pointer is pointing to with the pivot:

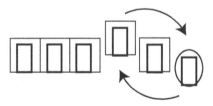

Although our array isn't completely sorted, we have successfully completed a partition. That is, since our pivot was the number 3, all numbers that are less than 3 are to its left, while all numbers greater than 3 are to its right. This also means that the 3 is *now in its correct place within the array*:

Below, we've implemented a SortableArray class in Ruby that includes a partition! method that partitions the array as we've described.

```ruby
class SortableArray

  attr_reader :array

  def initialize(array)
    @array = array
  end

  def partition!(left_pointer, right_pointer)

    # We always choose the right-most element as the pivot
    pivot_position = right_pointer
    pivot = @array[pivot_position]

    # We start the right pointer immediately to the left of the pivot
    right_pointer -= 1

    while true do

      while @array[left_pointer] < pivot do
        left_pointer += 1
      end

      while @array[right_pointer] > pivot do
        right_pointer -= 1
      end

      if left_pointer >= right_pointer
        break
      else
        swap(left_pointer, right_pointer)
      end

    end

    # As a final step, we swap the left pointer with the pivot itself
    swap(left_pointer, pivot_position)

    # We return the left_pointer for the sake of the quicksort method
    # which will appear later in this chapter
    return left_pointer
  end

  def swap(pointer_1, pointer_2)
    temp_value = @array[pointer_1]
    @array[pointer_1] = @array[pointer_2]
    @array[pointer_2] = temp_value
  end

end
```

Note that the partition! method accepts the starting points of the left and right pointers as parameters, and returns the end position of the left pointer once it's complete. This is all necessary to implement Quicksort, as we'll see next.

Quicksort

The Quicksort algorithm relies heavily on partitions. It works as follows:

1. Partition the array. The pivot is now in its proper place.

2. Treat the subarrays to the left and right of the pivot as their own arrays, and recursively repeat steps #1 and #2. That means that we'll partition each subarray, and end up with even smaller subarrays to the left and right of each subarray's pivot. We then partition those subarrays, and so on and so forth.

3. When we have a subarray that has zero or one elements, that is our base case and we do nothing.

Below is a quicksort! method that we can add to the preceding SortableArray class that would successfully complete Quicksort:

```
def quicksort!(left_index, right_index)
  #base case: the subarray has 0 or 1 elements
  if right_index - left_index <= 0
    return
  end

  # Partition the array and grab the position of the pivot
  pivot_position = partition!(left_index, right_index)

  # Recursively call this quicksort method on whatever is to the left of the pivot:
  quicksort!(left_index, pivot_position - 1)

  # Recursively call this quicksort method on whatever is to the right of the pivot:
  quicksort!(pivot_position + 1, right_index)
end
```

To see this in action, we'd run the following code:

```
array = [0, 5, 2, 1, 6, 3]
sortable_array = SortableArray.new(array)
sortable_array.quicksort!(0, array.length - 1)
p sortable_array.array
```

Let's return to our example. We began with the array of [0, 5, 2, 1, 6, 3] and ran a single partition on the entire array. Since Quicksort begins with such a partition, that means that we're already partly through the Quicksort process. We left off with:

As you can see, the value 3 was the pivot. Now that the pivot is in the correct place, we need to sort whatever is to the left and right of the pivot. Note that in our example, it just so happens that the numbers to the left of the pivot are already sorted, but the computer doesn't know that yet.

The next step after the partition is to treat everything to the left of the pivot as its own array and partition it.

We'll obscure the rest of the array for now, as we're not focusing on it at the moment:

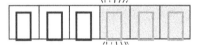

Now, of this [0, 1, 2] subarray, we'll make the right-most element the pivot. So that would be the number 2:

We'll establish our left and right pointers:

And now we're ready to partition this subarray. Let's continue from after Step #8, where we left off previously.

Step #9: We compare the left pointer (0) to the pivot (2). Since the 0 is less than the pivot, we continue moving the left pointer.

Step #10: We move the left pointer one cell to the right, and it now happens to be pointing to the same value that the right pointer is pointing to:

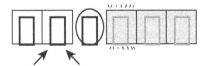

We compare the left pointer to the pivot. Since the value 1 is less than the pivot, we move on.

Step #11: We move the left pointer one cell to the right, which just happens to be the pivot:

(right pointer) (left pointer)

At this point, the left pointer is pointing to a value that is equal to the pivot (since it *is* the pivot!), and so the left pointer stops.

Step #12: Now, we activate the right pointer. However, since the right pointer's value (1) is less than the pivot, it stays still.

Since our left pointer has passed our right pointer, we're done moving pointers altogether in this partition.

Step #13: As a final step, we swap the pivot with the left pointer's value. Now, it just so happens that the left pointer is pointing to the pivot itself, so we swap the pivot with itself, which results in no change at all. At this point, the partition is complete and the pivot (2) is now in its correct spot:

We now have a subarray of [0, 1] to the left of the pivot (the 2) and no subarray to its right. The next step is to recursively partition the subarray to the pivot's left, which again is [0, 1]. We don't have to deal with any subarray to the right of the pivot since no such subarray exists.

Since all we'll focus on in the next step is the subarray [0, 1], we'll block out the rest of the array so it looks like this:

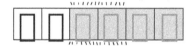

To partition the subarray [0, 1], we'll make the right-most element (the 1) the pivot. Where will we put the left and right pointers? Well, the left pointer will point to the 0, but the right pointer will also point to the 0, since we always start the right pointer out at one cell to the left of the pivot. This gives us this:

We're now ready to begin the partition.

Step #14: We compare the left pointer (0) with the pivot (1):

It's less than the pivot, so we move on.

Step #15: We shift the left pointer one cell to the right. It now points to the pivot:

Since the left pointer's value (1) is not lower than the pivot (since it *is* the pivot), the left pointer stops moving.

Step #16: We compare the right pointer with the pivot. Since it's pointing to a value that is less than the pivot, we don't move it anymore. Since the left pointer has passed the right pointer, we're done moving pointers for this partition. We're ready for the final step.

Step #17: We now swap the left pointer with the pivot. Again, in this case, the left pointer is actually pointing to the pivot itself, so the swap doesn't actually change anything. The pivot is now in its proper place, and we're done with this partition.

That leaves us with this:

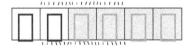

Next up, we need to partition the subarray to the left of the most recent pivot. In this case, that subarray is [0]—an array of just one element. An array of zero or one elements is our base case, so we don't do anything. The element is just considered to be in its proper place automatically. So now we've got:

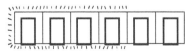

We started out by treating 3 as our pivot, and partitioning the subarray to its left ([0, 1, 2]). As promised, we now need to come back to partitioning the subarray to the right of the 3, which is [6, 5].

We'll obscure the [0, 1, 2, 3], since we've already sorted those, and now we're only focusing on the [6, 5]:

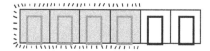

In the next partition, we'll treat the right-most element (the 5) as the pivot. That gives us:

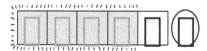

When setting up our next partition, our left and right pointers both end up pointing to the 6:

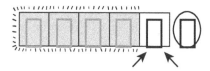

Step #18: We compare the left pointer (6) with the pivot (5). Since 6 is greater than the pivot, the left pointer doesn't move further.

Step #19: The right pointer is pointing to the 6 as well, so would theoretically move on to the next cell to the left. However, there are no more cells to the left of the 6, so the right pointer stops moving. Since the left pointer has reached the right pointer, we're done moving pointers altogether for this partition. That means we're ready for the final step.

Step #20: We swap the pivot with the value of the left pointer:

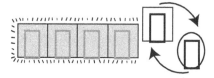

Our pivot (5) is now in its correct spot, leaving us with:

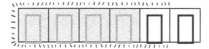

Next up, we technically need to recursively partition the subarray to the left and right of the [5, 6] subarray. Since there is no subarray to its left, that means that we only need to partition the subarray to the right. Since the subarray to the right of the 5 is a single element of [6], that's our base case and we do nothing—the 6 is automatically considered to be in its proper place:

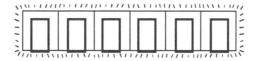

And we're done!

The Efficiency of Quicksort

To figure out the efficiency of Quicksort, let's first determine the efficiency of a partition. When breaking down the steps of a partition, we'll note that a partition consists of two types of steps:

- *Comparisons*: We compare each value to the pivot.
- *Swaps*: When appropriate, we swap the values being pointed to by the left and right pointers.

Each partition has at least N comparisons—that is, we compare each element of the array with the pivot. This is true because a partition always has the left and right pointers move through each cell until the left and right pointers reach each other.

The number of swaps depends on how the data is sorted. Each partition has at least one swap, and the most swaps that a partition can have would be N / 2, since even if we'd swap every possible value, we'd still be swapping the ones from the left half with the ones from the right half. See this diagram:

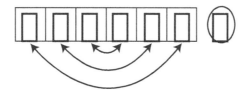

Now, for *randomly* sorted data, there would be roughly *half* of N / 2 swaps, or N / 4 swaps. So with N comparisons and N / 4, we'd say there are about 1.25N steps. In Big O Notation, we ignore constants, so we'd say that a partition runs in O(N) time.

So that's the efficiency of a single partition. But Quicksort involves many partitions on arrays and subarrays of different sizes, so we need to do a further analysis to determine the efficiency of Quicksort.

To visualize this more easily, here is a diagram depicting a typical Quicksort on an array of eight elements from a bird's-eye view. In particular, the diagram shows how many elements each partition acts upon. We've left out the actual numbers from the array since the exact numbers don't matter. Note that in the diagram, the active subarray is the group of cells that are not grayed out.

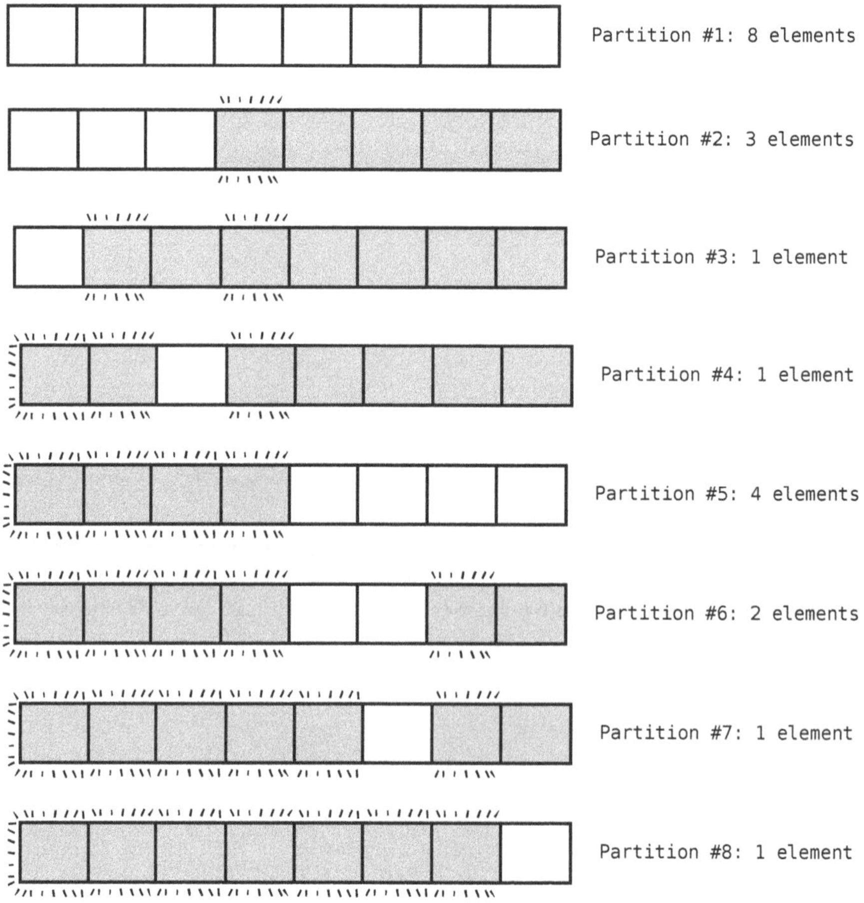

Partition #1: 8 elements

Partition #2: 3 elements

Partition #3: 1 element

Partition #4: 1 element

Partition #5: 4 elements

Partition #6: 2 elements

Partition #7: 1 element

Partition #8: 1 element

We can see that we have eight partitions, but each partition takes place on subarrays of various sizes. Now, where a subarray is just one element, that's our base case, and we don't perform any swaps or comparisons, so the significant partitions are the ones that take place on subarrays of two elements or more.

Since this example represents an average case, let's assume that each partition takes approximately 1.25N steps. So that gives us:

```
  8 elements * 1.25 steps = 10 steps
  3 elements * 1.25 steps = 3.75 steps
  4 elements * 1.25 steps = 5 steps
+ 2 elements * 1.25 steps = 2.5 steps
  _____
Total = Around 21 steps
```

If we do this analysis for arrays of various sizes, we'll see that for N elements, there are about N * log N steps. To get a feel for N * log N, see the following table:

N	log N	N * log N
4	2	8
8	3	24
16	4	64

In the preceding example, for an array of size 8, Quicksort took about 21 steps, which is about what 8 * log 8 is (24). This is the first time we're encountering such an algorithm, and in Big O, it's expressed as *O(N log N)*.

Now, it's not a coincidence that the number of steps in Quicksort just happens to align with N * log N. If we think about Quicksort more broadly, we can see *why* it is this way:

When we begin Quicksort, we partition the entire array. Assuming that the pivot ends up somewhere in the middle of the array—which is what happens in the average case—we then break up the two halves of the array and partition each half. We then take each of *those* halves, break them up into quarters, and then partition each quarter. And so on and so forth.

In essence, we keep on breaking down each subarray into halves until we reach subarrays with elements of 1. How many times can we break up an array in this way? For an array of N, we can break it down log N times. That is, an array of 8 can be broken in half three times until we reach elements of 1. You're already familiar with this concept from binary search.

Now, for each round in which we break down each subarray into two halves, we then partition all the elements from the original array over again. See this diagram:

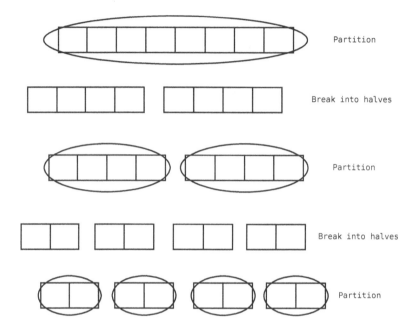

Since we can break up the original array into equal subarrays log N times, and for each time we break them up, we must run partitions on all N cells from the original array, we end up with about N * log N steps.

For many other algorithms we've encountered, the best case was one where the array was already sorted. When it comes to Quicksort, however, the best-case scenario is one in which the pivot always ends up smack in the middle of the subarray after the partition.

Worst-Case Scenario

The worst-case scenario for Quicksort is one in which the pivot always ends up on one side of the subarray instead of the middle. This can happen in several cases, including where the array is in perfect ascending or descending order. The visualization for this process is shown on page 127.

While, in this case, each partition only involves one swap, we lose out because of the many comparisons. In the first example, when the pivot always ended up towards the middle, each partition after the first one was conducted on relatively small subarrays (the largest subarray had a size of 4). In this

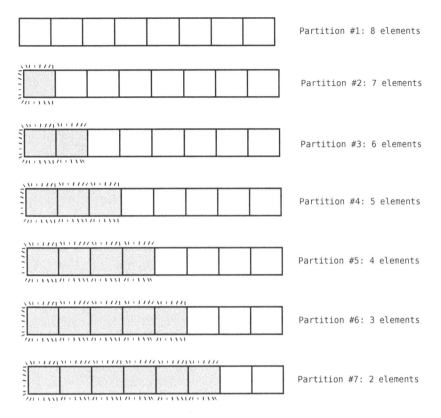

Partition #1: 8 elements
Partition #2: 7 elements
Partition #3: 6 elements
Partition #4: 5 elements
Partition #5: 4 elements
Partition #6: 3 elements
Partition #7: 2 elements

example, however, the first five partitions take place on subarrays of size 4 or greater. And each of these partitions have as many comparisons as there are elements in the subarray.

So in this worst-case scenario, we have partitions of 8 + 7 + 6 + 5 + 4 + 3 + 2 elements, which would yield a total of 35 comparisons.

To put this a little more formulaically, we'd say that for N elements, there are N + (N - 1) + (N - 2) + (N - 3) ... + 2 steps. This always comes out to be about $N^2 / 2$ steps, as you can see in the figures on page 128.

Since Big O ignores constants, we'd say that in a worst-case scenario, Quicksort has an efficiency of $O(N^2)$.

Now that we've got Quicksort down, let's compare it with Insertion Sort:

	Best case	Average case	Worst case
Insertion Sort	$O(N)$	$O(N^2)$	$O(N^2)$
Quicksort	$O(N \log N)$	$O(N \log N)$	$O(N^2)$

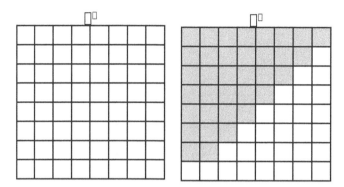

We can see that they have identical worst-case scenarios, and that Insertion Sort is actually faster than Quicksort for a best-case scenario. However, the reason why Quicksort is so much more superior than Insertion Sort is because of the average scenario—which, again, is what happens most of the time. For average cases, Insertion Sort takes a whopping $O(N^2)$, while Quicksort is much faster at $O(N \log N)$.

The following graph depicts various efficiencies side by side:

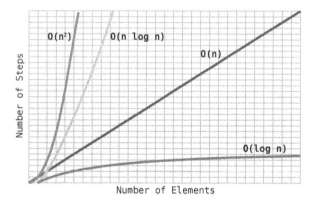

Because of Quicksort's superiority in average circumstances, many languages implement their own sorting functions by using Quicksort under the hood. Because of this, it's unlikely that you'll be implementing Quicksort yourself. However, there is a very similar algorithm that can come in handy for practical cases—and it's called Quickselect.

Quickselect

Let's say that you have an array in random order, and you do not need to sort it, but you do want to know the tenth-lowest value in the array, or the fifth-highest. This can be useful if we had a lot of test grades and wanted to know what the 25th percentile was, or if we wanted to find the median grade.

The obvious way to solve this would be to sort the entire array and then jump to the appropriate cell.

Even were we to use a fast sorting algorithm like Quicksort, this algorithm would take at least O(N log N) for average cases, and while that isn't bad, we can do even better with a brilliant little algorithm known as *Quickselect*. Quickselect relies on partitioning just like Quicksort, and can be thought of as a hybrid of Quicksort and binary search.

As we've seen earlier in this chapter, after a partition, the pivot value ends up in the appropriate spot in the array. Quickselect takes advantage of this in the following way:

Let's say that we have an array of eight values, and we want to find the second-to-lowest value within the array.

First, we partition the entire array:

After the partition, the pivot will hopefully end up somewhere towards the middle of the array:

This pivot is now in its correct spot, and since it's in the fifth cell, we now know which value is the fifth-lowest value within the array. Now, we're looking for the second-lowest value. But we also now know that the second-lowest value is somewhere to the left of the pivot. We can now ignore everything to the right of the pivot, and focus on the left subarray. It is in this respect that Quickselect is similar to binary search: we keep dividing the array in half, and focus on the half in which we know the value we're seeking for will be found.

Next, we partition the subarray to the left of the pivot:

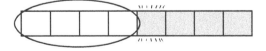

Let's say that the new pivot of this subarray ends up the third cell:

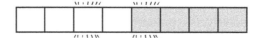

We now know that the value in the third cell is in its correct spot, meaning that it's the third-to-lowest value in the array. By definition, then, the second-to-lowest value will be somewhere to its left. We can now partition the subarray to the left of the third cell:

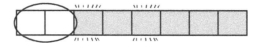

After this next partition, the lowest and second-lowest values will end up in their correct spots within the array:

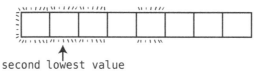

second lowest value

We can then grab the value from the second cell and know with confidence that it's the second-lowest value in the entire array. One of the beautiful things about Quickselect is that we can find the correct value *without having to sort the entire array*.

With Quicksort, for every time we cut the array in half, we need to re-partition every single cell from the original array, giving us O(N log N). With Quickselect, on the other hand, for every time we cut the array in half, we only need to partition the one half that we care about—the half in which we know our value is to be found.

When analyzing the efficiency of Quickselect, we'll see that it's O(N) for average scenarios. Recall that each partition takes about N steps for the subarray it's run upon. Thus, for Quickselect on an array of eight elements, we run partitions three times: on an array of eight elements, on a subarray of four elements, and on a subarray of two elements. This is a total of 8 + 4 + 2 = 14 steps. So an array of 8 elements yields roughly 14 steps.

For an array of 64 elements, we run 64 + 32 + 16 + 8 + 4 + 2 = 126 steps. For 128 elements, we would have 254 steps. And for 256 elements, we would have 510 steps.

If we were to formulate this, we would say that for N elements, we would have N + (N/2) + (N/4) + (N/8) + ... 2 steps. This always turns out to be roughly

2N steps. Since Big O ignores constants, we would say that Quickselect has an efficiency of O(N).

Below, we've implemented a quickselect! method that can be dropped into the SortableArray class described previously. You'll note that it's very similar to the quicksort! method:

```ruby
def quickselect!(kth_lowest_value, left_index, right_index)
  # If we reach a base case - that is, that the subarray has one cell,
  # we know we've found the value we're looking for
  if right_index - left_index <= 0
    return @array[left_index]
  end

  # Partition the array and grab the position of the pivot
  pivot_position = partition!(left_index, right_index)

  if kth_lowest_value < pivot_position
    quickselect!(kth_lowest_value, left_index, pivot_position - 1)
  elsif kth_lowest_value > pivot_position
    quickselect!(kth_lowest_value, pivot_position + 1, right_index)
  else # kth_lowest_value == pivot_position
    # if after the partition, the pivot position is in the same spot
    # as the kth lowest value, we've found the value we're looking for
    return @array[pivot_position]
  end
end
```

If you wanted to find the second-to-lowest value of an unsorted array, you'd run the following code:

```ruby
array = [0, 50, 20, 10, 60, 30]
sortable_array = SortableArray.new(array)
p sortable_array.quickselect!(1, 0, array.length - 1)
```

The first argument of the quickselect! method accepts the position that you're looking for, starting at index 0. We've put in a 1 to represent the second-to-lowest value.

Wrapping Up

The Quicksort and Quickselect algorithms are recursive algorithms that present beautiful and efficient solutions to thorny problems. They're great examples of how a non-obvious but well-thought-out algorithm can boost performance.

Algorithms aren't the only things that are recursive. Data structures can be recursive as well. The data structures that we will encounter in the next few chapters—linked list, binary tree, and graph—use their recursive natures to allow for speedy data access in their own ingenious ways.

CHAPTER 11

Node-Based Data Structures

For the next several chapters, we're going to explore a variety of data structures that all build upon a single concept—the *node*. Node-based data structures offer new ways to organize and access data that provide a number of major performance advantages.

In this chapter, we're going to explore the linked list, which is the simplest node-based data structure and the foundation for the future chapters. You're also going to discover that linked lists seem almost identical to arrays, but come with their own set of trade-offs in efficiency that can give us a performance boost for certain situations.

Linked Lists

A *linked list* is a data structure that represents a list of items, just like an array. In fact, in any application in which you're using an array, you could probably use a linked list instead. Under the hood, however, linked lists are implemented differently and can have different performance in varying situations.

As mentioned in our first chapter, memory inside a computer exists as a giant set of cells in which bits of data are stored. When creating an array, your code finds a contiguous group of empty cells in memory and designates them to store data for your application as shown on page 134.

We also explained there that the computer has the ability to jump straight to any index within the array. If you wrote code that said, "Look up the value at index 4," your computer could locate that cell in a single step. Again, this is because your program knows which memory address the array starts at—say, memory address 1000—and therefore knows that if it wants to look up index 4, it should simply jump to memory address 1004.

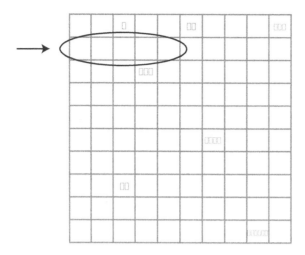

Linked lists, on the other hand, do not consist of a bunch of memory cells in a row. Instead, they consist of a bunch of memory cells that are *not* next to each other, but can be spread across many different cells across the computer's memory. These cells that are not adjacent to each other are known as *nodes*.

The big question is: if these nodes are not next to each other, how does the computer know which nodes are part of the same linked list?

This is the key to the linked list: in addition to the data stored within the node, *each node also stores the memory address of the next node in the linked list.*

This extra piece of data—this pointer to the next node's memory address—is known as a *link*. Here is a visual depiction of a linked list:

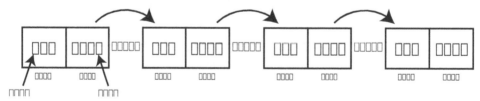

In this example, we have a linked list that contains four pieces of data: "a", "b", "c", and "d". However, it uses *eight* cells of memory to store this data, because each node consists of two memory cells. The first cell holds the actual data, while the second cell serves as a link that indicates where in memory the next node begins. The final node's link contains null since the linked list ends there.

For our code to work with a linked list, all it really needs to know is where in memory the first node begins. Since each node contains a link to the next node, if the application is given the first node of the linked list, it can piece together the rest of the list by following the link of the first node to the second node, and the link from the second node to the third node, and so on.

One advantage of a linked list over an array is that the program doesn't need to find a bunch of empty memory cells in a row to store its data. Instead, the program can store the data across many cells that are not necessarily adjacent to each other.

Implementing a Linked List

Let's implement a linked list using Ruby. We'll use two classes to implement this: `Node` and `LinkedList`. Let's create the `Node` class first:

```ruby
class Node
  attr_accessor :data, :next_node

  def initialize(data)
    @data = data
  end
end
```

The `Node` class has two attributes: `data` contains the value that the node is meant to hold, while `next_node` contains the link to the next node in the list. We can use this class as follows:

```ruby
node_1 = Node.new("once")
node_2 = Node.new("upon")
node_1.next_node = node_2

node_3 = Node.new("a")
node_2.next_node = node_3

node_4 = Node.new("time")
node_3.next_node = node_4
```

With this code, we've created a list of four nodes that serve as a list containing the strings "once", "upon", "a", and "time".

While we've been able to create this linked list with the `Node` class alone, we need an easy way to tell our program where the linked list begins. To do this, we'll create a `LinkedList` class as well. Here's the `LinkedList` class in its basic form:

```ruby
class LinkedList
  attr_accessor :first_node

  def initialize(first_node)
    @first_node = first_node
  end
end
```

With this class, we can tell our program about the linked list we created previously using the following code:

```
list = LinkedList.new(node_1)
```

This `LinkedList` acts as a handle on the linked list by pointing to its first node.

Now that we know what a linked list *is*, let's measure its performance against a typical array. We'll do this by analyzing the four classic operations: reading, searching, insertion, and deletion.

Reading

We noted previously that when reading a value from an array, the computer can jump to the appropriate cell in a single step, which is O(1). This is not the case, however, with a linked list.

If your program wanted to read the value at index 2 of a linked list, the computer could not look it up in one step, because it wouldn't immediately know where to find it in the computer's memory. After all, each node of a linked list could be *anywhere* in memory! Instead, all the program knows is the memory address of the *first* node of the linked list. To find index 2 (which is the third node), the program must begin looking up index 0 of the linked list, and then follow the link at index 0 to index 1. It must then again follow the link at index 1 to index 2, and finally inspect the value at index 2.

Let's implement this operation inside our `LinkedList` class:

```ruby
class LinkedList

  attr_accessor :first_node

  def initialize(first_node)
    @first_node = first_node
  end

  def read(index)
    # We begin at the first node of the list:
    current_node = first_node
    current_index = 0

    while current_index < index do
      # We keep following the links of each node until we get to the
      # index we're looking for:
      current_node = current_node.next_node
      current_index += 1

      # If we're past the end of the list, that means the
      # value cannot be found in the list, so return nil:
      return nil unless current_node
    end
```

```
      return current_node.data
  end
end
```

We can then look up a particular index within the list with:

`list.read(3)`

If we ask a computer to look up the value at a particular index, the worst-case scenario would be if we're looking up the last index in our list. In such a case, the computer will take N steps to look up this index, since it needs to start at the first node of the list and follow each link until it reaches the final node. Since Big O Notation expresses the worst-case scenario unless explicitly stated otherwise, we would say that reading a linked list has an efficiency of O(N). This is a significant disadvantage in comparison with arrays in which reads are O(1).

Searching

Arrays and linked lists have the same efficiency for search. Remember, searching is looking for a particular piece of data within the list and getting its index. With both arrays and linked lists, the program needs to start at the first cell and look through each and every cell until it finds the value it's searching for. In a worst-case scenario—where the value we're searching for is in the final cell or not in the list altogether—it would take O(N) steps.

Here's how we'd implement the search operation:

```
class LinkedList

  attr_accessor :first_node

  # rest of code omitted here...

  def index_of(value)
    # We begin at the first node of the list:
    current_node = first_node
    current_index = 0

    begin
      # If we find the data we're looking for, we return it:
      if current_node.data == value
        return current_index
      end

      # Otherwise, we move on the next node:
      current_node = current_node.next_node
      current_index += 1
    end while current_node
```

```
    # If we get through the entire list, without finding the
    # data, we return nil:
    return nil
  end
end
```

We can then search for any value within the list using:

```
list.index_of("time")
```

Insertion

Insertion is one operation in which linked lists can have a distinct advantage over arrays in certain situations. Recall that the worst-case scenario for insertion into an array is when the program inserts data into index 0, because it then has to shift the rest of the data one cell to the right, which ends up yielding an efficiency of O(N). With linked lists, however, insertion at the beginning of the list takes just one step—which is O(1). Let's see why.

Say that we had the following linked list:

If we wanted to add "yellow" to the beginning of the list, all we would have to do is create a new node and have it link to the node containing "blue":

In contrast with the array, therefore, a linked list provides the flexibility of inserting data to the front of the list without requiring the shifting of any other data.

The truth is that, theoretically, inserting data *anywhere* within a linked list takes just one step, but there's one gotcha. Let's say that we now have this linked list:

And say that we want to insert "purple" at index 2 (which would be between "blue" and "green"). The actual insertion takes one step, since we create the new purple node and simply modify the blue node to point to it as shown on page 139.

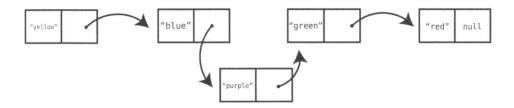

However, for the computer to do this, it first needs to *find* the node at index 1 ("blue") so it can modify its link to point to the newly created node. As we've seen, though, reading from a linked list already takes O(N). Let's see this in action.

We know that we want to add a new node after index 1. So the computer needs to find index 1 of the list. To do this, we must start at the beginning of the list:

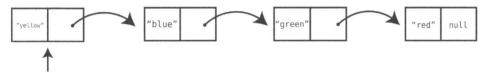

We then access the next node by following the first link:

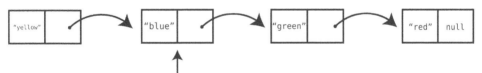

Now that we've found index 1, we can finally add the new node:

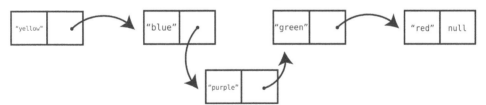

In this case, adding "purple" took three steps. If we were to add it to the *end* of our list, it would take five steps: four steps to find index 3, and one step to insert the new node.

Therefore, inserting into the middle of a linked list takes O(N), just as it does for an array.

Interestingly, our analysis shows that the best- and worst-case scenarios for arrays and linked lists are the opposite of one another. That is, inserting at the beginning is great for linked lists, but terrible for arrays. And inserting at the end is an array's best-case scenario, but the worst case when it comes to a linked list. The following table breaks this all down:

Scenario	Array	Linked list
Insert at beginning	Worst case	Best case
Insert at middle	Average case	Average case
Insert at end	Best case	Worst case

Here's how we can implement the insert operation in our LinkedList class:

```ruby
class LinkedList
  attr_accessor :first_node

  # rest of code omitted here...

  def insert_at_index(index, value)
    # We create the new node:
    new_node = Node.new(value)

    # If we are inserting at beginning of list:
    if index == 0
      # Have our new node link to what was the first node
      new_node.next_node = first_node
      # Establish that the first node will now be our new node:
      self.first_node = new_node
    else
      current_node = first_node
      current_index = 0

      # First, we find the index immediately before where the
      # new node will go:
      while current_index < (index - 1) do
        current_node = current_node.next_node
        current_index += 1
      end

      # We have the new node link to the next node
      new_node.next_node = current_node.next_node

      # We modify the link of the previous node to point to
      # our new node:
      current_node.next_node = new_node
    end
  end
end
```

Deletion

Deletion is very similar to insertion in terms of efficiency. To delete a node from the beginning of a linked list, all we need to do is perform one step: we change the first_node of the linked list to now point to the second node.

Let's return to our example of the linked list containing the values "once", "upon", "a", and "time". If we wanted to delete the value "once", we would simply change the linked list to begin at "upon":

```
list.first_node = node_2
```

Contrast this with an array in which deleting the first element means shifting all remaining data one cell to the left, an efficiency of O(N).

When it comes to deleting the *final* node of a linked list, the actual deletion takes one step—we just take the second-to-last node and make its link null. However, it takes N steps to first get to the second-to-last node, since we need to start at the beginning of the list and follow the links until we reach it.

The following table contrasts the various scenarios of deletion for both arrays and linked lists. Note how it's identical to insertion:

Situation	Array	Linked list
Delete at beginning	Worst case	Best case
Delete at middle	Average case	Average case
Delete at end	Best case	Worst case

To delete from the middle of the list, the computer must modify the link of the preceding node. The following example will make this clear.

Let's say that we want to delete the value at index 2 ("purple") from the previous linked list. The computer finds index 1 and switches its link to point to the "green" node:

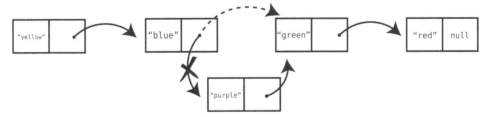

Here's what the delete operation would look like in our class:

```
class LinkedList

  attr_accessor :first_node

  # rest of code omitted here...

  def delete_at_index(index)
    # If we are deleting the first node:
    if index == 0
      # Simply set the first node to be what is currently the second node:
      self.first_node = first_node.next_node
```

```
    else
      current_node = first_node
      current_index = 0

      # First, we find the node immediately before the one we
      # want to delete and call it current_node:
      while current_index < (index - 1) do
        current_node = current_node.next_node
        current_index += 1
      end

      # We find the node that comes after the one we're deleting:
      node_after_deleted_node = current_node.next_node.next_node

      # We change the link of the current_node to point to the
      # node_after_deleted_node, leaving the node we want
      # to delete out of the list:
      current_node.next_node = node_after_deleted_node
    end
  end
end
```

After our analysis, it emerges that the comparison of linked lists and arrays breaks down as follows:

Operation	Array	Linked list
Reading	O(1)	O(N)
Search	O(N)	O(N)
Insertion	O(N) (O(1) at end)	O(N) (O(1) at beginning)
Deletion	O(N) (O(1) at end)	O(N) (O(1) at beginning)

Search, insertion, and deletion seem to be a wash, and reading from an array is much faster than reading from a linked list. If so, why would one ever want to use a linked list?

Linked Lists in Action

One case where linked lists shine is when we examine a single list and delete many elements from it. Let's say, for example, that we're building an application that combs through lists of email addresses and removes any email address that has an invalid format. Our algorithm inspects each and every email address one at a time, and uses a regular expression (a specific pattern for identifying certain types of data) to determine whether the email address is invalid. If it's invalid, it removes it from the list.

No matter whether the list is an array or a linked list, we need to comb through the entire list one element at a time to inspect each value, which would take

N steps. However, let's examine what happens when we actually delete email addresses.

With an array, each time we delete an email address, we need another O(N) steps to shift the remaining data to the left to close the gap. All this shifting will happen before we can even inspect the next email address.

So besides the N steps of reading each email address, we need another N steps multiplied by invalid email addresses to account for deletion of invalid email addresses.

Let's assume that one in ten email addresses are invalid. If we had a list of 1,000 email addresses, there would be approximately 100 invalid ones, and our algorithm would take 1,000 steps for reading plus about 100,000 steps for deletion (100 invalid email addresses * N).

With a linked list, however, as we comb through the list, each deletion takes just one step, as we can simply change a node's link to point to the appropriate node and move on. For our 1,000 emails, then, our algorithm would take just 1,100 steps, as there are 1,000 reading steps, and 100 deletion steps.

Doubly Linked Lists

Another interesting application of a linked list is that it can be used as the underlying data structure behind a queue. We covered queues in *Crafting Elegant Code*, and you'll recall that they are lists of items in which data can only be inserted at the end and removed from the beginning. Back then, we used an array as the basis for the queue, explaining that the queue is simply an array with special constraints. However, we can also use a linked list as the basis for a queue, assuming that we enforce the same constraints of only inserting data at the end and removing data from the beginning. Does using a linked list instead of an array have any advantages? Let's analyze this.

Again, the queue inserts data at the end of the list. As we discussed earlier in this chapter, arrays are superior when it comes to inserting data, since we're able to do so at an efficiency of O(1). Linked lists, on the other hand, insert data at O(N). So when it comes to insertion, the array makes for a better choice than a linked list.

When it comes to deleting data from a queue, though, linked lists are faster, since they are O(1) compared to arrays, which delete data from the beginning at O(N).

Based on this analysis, it would seem that it doesn't matter whether we use an array or a linked list, as we'd end up with one major operation that is O(1) and another that is O(N). For arrays, insertion is O(1) and deletion is O(N), and for linked lists, insertion is O(N) and deletion is O(1).

However, if we use a special variant of a linked list called the *doubly linked list*, we'd be able to insert *and* delete data from a queue at O(1).

A doubly linked list is like a linked list, except that each node has *two* links—one that points to the next node, and one that points to the *preceding* node. In addition, the doubly linked list keeps track of both the first and last nodes.

Here's what a doubly linked list looks like:

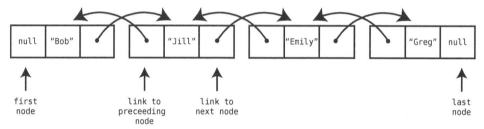

In code, the core of a doubly linked list would look like this:

```ruby
class Node
  attr_accessor :data, :next_node, :previous_node

  def initialize(data)
    @data = data
  end
end

class DoublyLinkedList
  attr_accessor :first_node, :last_node

  def initialize(first_node=nil, last_node=nil)
    @first_node = first_node
    @last_node = last_node
  end
end
```

Since a doubly linked list always knows where both its first and last nodes are, we can access each of them in a single step, or O(1). Similarly, we can insert data at the end of a doubly linked list in one step by doing the following:

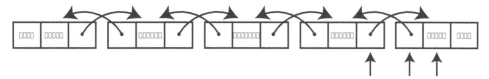

We create a new node ("Sue") and have its previous_node point to the last_node of the linked list ("Greg"). Then, we change the next_node of the last_node ("Greg") to point to this new node ("Sue"). Finally, we declare the new node ("Sue") to be the last_node of the linked list.

Here's the implementation of a new insert_at_end method available to doubly linked lists:

```ruby
class DoublyLinkedList

  attr_accessor :first_node, :last_node

  def initialize(first_node=nil, last_node=nil)
    @first_node = first_node
    @last_node = last_node
  end

  def insert_at_end(value)
    new_node = Node.new(value)

    # If there are no elements yet in the linked list:
    if !first_node
      @first_node = new_node
      @last_node = new_node
    else
      new_node.previous_node = @last_node
      @last_node.next_node = new_node
      @last_node = new_node
    end
  end

end
```

Because doubly linked lists have immediate access to both the front and end of the list, they can insert data on either side at O(1) as well as delete data on either side at O(1). Since doubly linked lists can insert data at the end in O(1) time and delete data from the front in O(1) time, they make the perfect underlying data structure for a queue.

Here's a complete example of a queue that is built upon a doubly linked list:

```ruby
class Node

  attr_accessor :data, :next_node, :previous_node

  def initialize(data)
    @data = data
  end
```

```ruby
  end

class DoublyLinkedList
  attr_accessor :first_node, :last_node

  def initialize(first_node=nil, last_node=nil)
    @first_node = first_node
    @last_node = last_node
  end

  def insert_at_end(value)
    new_node = Node.new(value)

    # If there are no elements yet in the linked list:
    if !first_node
      @first_node = new_node
      @last_node = new_node
    else
      new_node.previous_node = @last_node
      @last_node.next_node = new_node
      @last_node = new_node
    end
  end

  def remove_from_front
    removed_node = @first_node
    @first_node = @first_node.next_node
    return removed_node
  end
end

class Queue
  attr_accessor :queue

  def initialize
    @queue = DoublyLinkedList.new
  end

  def enque(value)
    @queue.insert_at_end(value)
  end

  def deque
    removed_node = @queue.remove_from_front
    return removed_node.data
  end

  def tail
    return @queue.last_node.data
  end
end
```

Wrapping Up

Your current application may not require a queue, and your queue may work just fine even if it's built upon an array rather than a doubly linked list. However, you are learning that you have choices—and you are learning how to make the right choice at the right time.

You've learned how linked lists can be useful for boosting performance in certain situations. In the next chapters, we're going to learn about more complex node-based data structures that are more common and used in many everyday circumstances to make applications run at greater efficiency.

CHAPTER 12

Speeding Up All the Things with Binary Trees

In *Why Algorithms Matter*, we covered the concept of binary search, and demonstrated that if we have an ordered array, we can use binary search to locate any value in O(log N) time. Thus, ordered arrays are awesome.

However, there is a problem with ordered arrays.

When it comes to insertions and deletions, ordered arrays are slow. Whenever a value is inserted into an ordered array, we first shift all items greater than that value one cell to the right. And when a value is deleted from an ordered array, we shift all greater values one cell to the left. This takes N steps in a worst-case scenario (inserting into or deleting from the first value of the array), and N / 2 steps on average. Either way, it's O(N), and O(N) is relatively slow.

Now, we learned in *Blazing Fast Lookup with Hash Tables* that hash tables are O(1) for search, insertion, and deletion, but they have one significant disadvantage: they do not maintain order.

But what do we do if we want a data structure that maintains order, and also has fast search, insertion, and deletion? Neither an ordered array nor a hash table is ideal.

Enter the binary tree.

Binary Trees

We were introduced to node-based data structures in the previous chapter using linked lists. In a simple linked list, each node contains a link that connects this node to a single other node. A *tree* is also a node-based data structure, but within a tree, each node can have links to *multiple* nodes.

Here is a visualization of a simple tree:

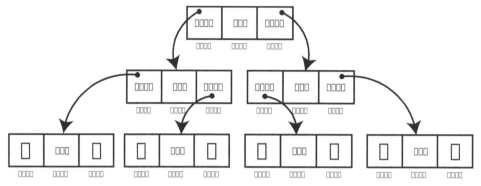

In this example, each node has links that lead to two other nodes. For the sake of simplicity, we can represent this tree visually without showing all the actual links:

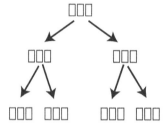

Trees come with their own unique nomenclature:

- The uppermost node (in our example, the "j") is called the *root*. Yes, in our picture the root is at the top of the tree; deal with it.

- In our example, we'd say that the "j" is a *parent* to "m" and "b", which are in turn *children* of "j". The "m" is a parent of "q" and "z", which are in turn children of "m".

- Trees are said to have *levels*. The preceding tree has three levels:

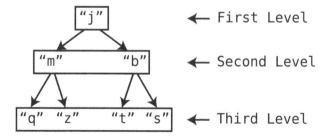

There are many different kinds of tree-based data structures, but in this chapter, we'll be focusing on a particular tree known as a binary tree. A *binary tree* is a tree that abides by the following rules:

- Each node has either zero, one, or two children.
- If a node has two children, it must have one child that has a lesser value than the parent, and one child that has a greater value than the parent.

Here's an example of a binary tree, in which the values are numbers:

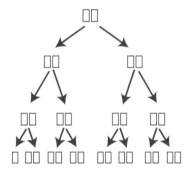

Note that each node has one child with a lesser value than itself, which is depicted using a left arrow, and one child with a greater value than itself, which is depicted using a right arrow.

While the following example is a tree, it is not a *binary* tree:

It is not a valid binary tree because both children of the parent node have values less than the parent itself.

The implementation of a tree node in Python might look something like this:

```python
class TreeNode:
    def __init__(self,val,left=None,right=None):
        self.value = val
        self.leftChild = left
        self.rightChild = right
```

We can then build a simple tree like this:

```python
node = TreeNode(1)
node2 = TreeNode(10)
root = TreeNode(5, node, node2)
```

Because of the unique structure of a binary tree, we can search for any value within it very, very quickly, as we'll now see.

Searching

Here again is a binary tree:

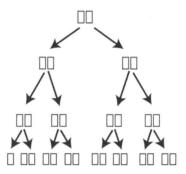

The algorithm for searching within a binary tree begins at the root node:

1. Inspect the value at the node.

2. If we've found the value we're looking for, great!

3. If the value we're looking for is less than the current node, search for it in its left subtree.

4. If the value we're looking for is greater than the current node, search for it in its right subtree.

Here's a simple, recursive implementation for this search in Python:

```python
def search(value, node):
    # Base case: If the node is nonexistent
    # or we've found the value we're looking for:
    if node is None or node.value == value:
        return node

    # If the value is less than the current node, perform
    # search on the left child:
    elif value < node.value:
        return search(value, node.leftChild)

    # If the value is less than the current node, perform
    # search on the right child:
    else:  # value > node.value
        return search(value, node.rightChild)
```

Say we wanted to search for the 61. Let's see how many steps it would take by walking through this visually.

When searching a tree, we must always begin at the root:

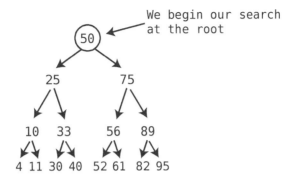

Next, the computer asks itself: is the number we're searching for (61) greater or less than the value of this node? If the number we're looking for is less than the current node, look for it in the left child. If it's greater than the current node, look for it in the right child.

In this example, since 61 is greater than 50, we know it must be somewhere to the right, so we search the right child:

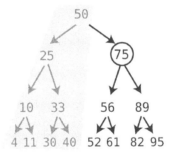

"Are you my mother?" asks the algorithm. Since the 75 is not the 61 we're looking for, we need to move down to the next level. And since 61 is less than 75, we'll check the left child, since the 61 could only be there:

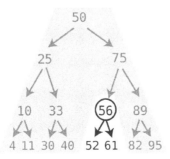

Since 61 is greater than 56, we search for it in the right child of the 56:

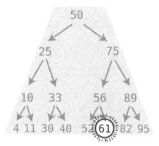

In this example, it took us four steps to find our desired value.

More generally, we'd say that searching in a binary tree is O(log N). This is because each step we take eliminates half of the remaining possible values in which our value can be stored. (We'll see soon, though, that this is only for a perfectly balanced binary tree, which is a best-case scenario.)

Compare this with binary search, another O(log N) algorithm, in which each number we try also eliminates half of the remaining possible values. In this regard, then, searching a binary tree has the same efficiency as binary search within an ordered array.

Where binary trees really shine over ordered arrays, though, is with insertion.

Insertion

To discover the algorithm for inserting a new value into a binary tree, let's work with an example. Say that we want to insert the number 45 into the previous tree.

The first thing we'd have to do is find the correct node to attach the 45 to. To begin our search, we start at the root:

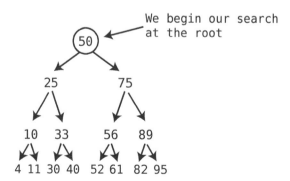

Since 45 is less than 50, we drill down to the left child:

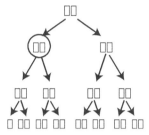

Since 45 is greater than 25, we must inspect the right child:

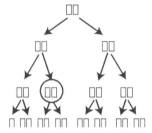

45 is greater than 33, so we check the 33's right child:

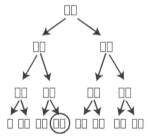

At this point, we've reached a node that has no children, so we have nowhere to go. This means we're ready to perform our insertion.

Since 45 is greater than 40, we insert it as a right child node of the 40:

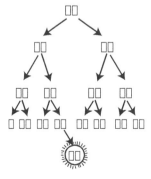

In this example, insertion took five steps, consisting of four search steps, and one insertion step. Insertion always takes just one extra step beyond a search, which means that insertion takes log N + 1 steps, which is O(log N) as Big O ignores constants.

In an ordered array, by contrast, insertion takes O(N), because in addition to search, we must shift a lot of data to the right to make room for the value that we're inserting.

This is what makes binary trees so efficient. While ordered arrays have O(log N) search and O(N) insertion, binary trees have O(log N) *and* O(log N) insertion. This becomes critical in an application where you anticipate a lot of changes to your data.

Here's a Python implementation of inserting a value into a binary tree. Like the search function, it is recursive:

```python
def insert(value, node):
    if value < node.value:

        # If the left child does not exist, we want to insert
        # the value as the left child:
        if node.leftChild is None:
            node.leftChild = TreeNode(value)
        else:
            insert(value, node.leftChild)

    elif value > node.value:

        # If the right child does not exist, we want to insert
        # the value as the right child:
        if node.rightChild is None:
            node.rightChild = TreeNode(value)
        else:
            insert(value, node.rightChild)
```

It is important to note that only when creating a tree out of randomly sorted data do trees usually wind up well-balanced. However, if we insert *sorted* data into a tree, it can become imbalanced and less efficient. For example, if we were to insert the following data in this order—1, 2, 3, 4, 5—we'd end up with a tree that looks like this:

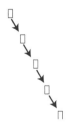

Searching for the 5 within this tree would take O(N).

However, if we inserted the same data in the following order—3, 2, 4, 1, 5—the tree would be evenly balanced:

Because of this, if you ever wanted to convert an ordered array into a binary tree, you'd better first randomize the order of the data.

It emerges that in a worst-case scenario, where a tree is completely *imbalanced*, search is O(N). In a best-case scenario, where it is perfectly balanced, search is O(log N). In the typical scenario, in which data is inserted in random order, a tree will be pretty well balanced and search will take about O(log N).

Deletion

Deletion is the least straightforward operation within a binary tree, and requires some careful maneuvering. Let's say that we want to delete the 4 from this binary tree:

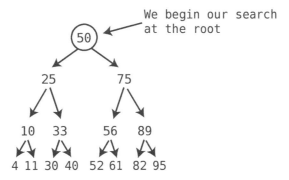

First, we perform a search to first find the 4, and then we can just delete it one step:

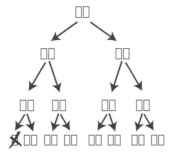

While that was simple, let's say we now want to delete the 10 as well. If we delete the 10

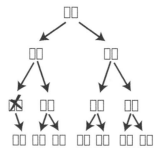

we end up with an 11 that isn't connected to the tree anymore. And we can't have that, because we'd lose the 11 forever. However, there's a solution to this problem: we'll plug the 11 into where the 10 used to be:

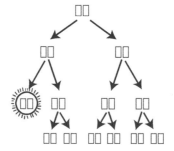

So far, our deletion algorithm follows these rules:

- If the node being deleted has no children, simply delete it.
- If the node being deleted has one child, delete it and plug the child into the spot where the deleted node was.

Deleting a node that has two children is the most complex scenario. Let's say that we wanted to delete the 56:

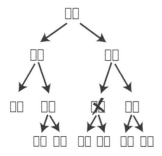

What are we going to do with the 52 and 61? We cannot put *both* of them where the 56 was. That is where the next rule of the deletion algorithm comes into play:

- When deleting a node with two children, replace the deleted node with the *successor* node. The successor node is the child node whose value is the *least of all values that are greater than the deleted node*.

That was a tricky sentence. In other words, the successor node is the next number up from the deleted value. In this case, the next number up among the descendants of 56 is 61. So we replace the 56 with the 61:

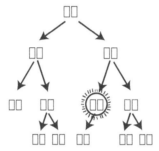

How does the computer find the successor value? There is an algorithm for that:

Visit the right child of the deleted value, and then keep on visiting the left child of each subsequent child until there are no more left children. The bottom value is the successor node.

Let's see this again in action in a more complex example. Let's delete the root node:

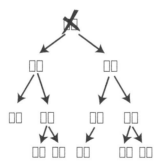

We now need to plug the successor value into the root node. Let's find the successor value.

To do this, we first visit the right child, and then keep descending leftward until we reach a node that doesn't have a left child:

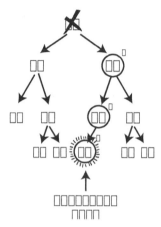

Now that we've found the successor node—the 52—we plug it into the node that we deleted:

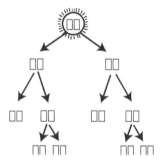

And we're done!

However, there is one case that we haven't accounted for yet, and that's where the successor node has a right child of its own. Let's recreate the preceding tree, but add a right child to the 52:

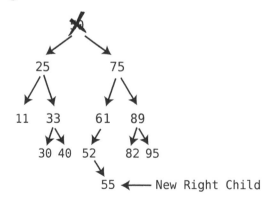

In this case, we can't simply plug the successor node—the 52—into the root, since we'd leave its child of 55 hanging. Because of this, there's one more rule to our deletion algorithm:

- If the successor node has a right child, after plugging the successor into the spot of the deleted node, take the right child of the successor node and turn it into the *left child of the parent of the successor node.*

Let's walk through the steps.

First, we plug the successor node into the root:

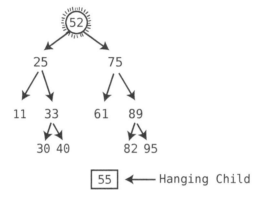

At this point, the 55 is left dangling. Next, we turn the 55 into the left child of what was the parent of the successor node. In this case, 61 was the parent of the successor node, so we make the 55 the left child of the 61:

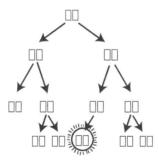

And now we're *really* done.

Pulling all the steps together, the algorithm for deletion from a binary tree is:

- If the node being deleted has no children, simply delete it.
- If the node being deleted has one child, delete it and plug the child into the spot where the deleted node was.

- When deleting a node with two children, replace the deleted node with the *successor* node. The successor node is the child node whose value is the least of all values that are *greater* than the deleted node.
  - If the successor node has a right child, after plugging the successor node into the spot of the deleted node, take the right child of the successor node and turn it into the *left child of the parent of the successor node.*

Here's a recursive Python implementation of deletion from a binary tree. It's a little tricky at first, so we've sprinkled comments liberally:

```python
def delete(valueToDelete, node):

    # The base case is when we've hit the bottom of the tree,
    # and the parent node has no children:
    if node is None:
        return None

    # If the value we're deleting is less or greater than the current node,
    # we set the left or right child respectively to be
    # the return value of a recursive call of this very method on the current
    # node's left or right subtree.
    elif valueToDelete < node.value:
        node.leftChild = delete(valueToDelete, node.leftChild)
        # We return the current node (and its subtree if existent) to
        # be used as the new value of its parent's left or right child:
        return node
    elif valueToDelete > node.value:
        node.rightChild = delete(valueToDelete, node.rightChild)
        return node

    # If the current node is the one we want to delete:
    elif valueToDelete == node.value:

        # If the current node has no left child, we delete it by
        # returning its right child (and its subtree if existent)
        # to be its parent's new subtree:
        if node.leftChild is None:
            return node.rightChild

            # (If the current node has no left OR right child, this ends up
            # being None as per the first line of code in this function.)

        elif node.rightChild is None:
            return node.leftChild

        # If the current node has two children, we delete the current node
        # by calling the lift function (below), which changes the current node's
        # value to the value of its successor node:
        else:
            node.rightChild = lift(node.rightChild, node)
            return node
```

```
def lift(node, nodeToDelete):

    # If the current node of this function has a left child,
    # we recursively call this function to continue down
    # the left subtree to find the successor node.
    if node.leftChild:
        node.leftChild = lift(node.leftChild, nodeToDelete)
        return node
    # If the current node has no left child, that means the current node
    # of this function is the successor node, and we take its value
    # and make it the new value of the node that we're deleting:
    else:
        nodeToDelete.value = node.value
        # We return the successor node's right child to be now used
        # as its parent's left child:
        return node.rightChild
```

Like search and insertion, deleting from trees is also typically O(log N). This is because deletion requires a search plus a few extra steps to deal with any hanging children. Contrast this with deleting a value from an ordered array, which takes O(N) due to shifting elements to the left to close the gap of the deleted value.

Binary Trees in Action

We've seen that binary trees boast efficiencies of O(log N) for search, insertion, and deletion, making it an efficient choice for scenarios in which we need to store and manipulate ordered data. This is particularly true if we will be modifying the data often, because while ordered arrays are just as fast as binary trees when searching data, binary trees are significantly faster when it comes to inserting and deleting data.

For example, let's say that we're creating an application that maintains a list of book titles. We'd want our application to have the following functionality:

- Our program should be able to print out the list of book titles in alphabetical order.
- Our program should allow for constant changes to the list.
- Our program should allow the user to search for a title within the list.

If we didn't anticipate that our booklist would be changing that often, an ordered array would be a suitable data structure to contain our data. However, we're building an app that should be able to handle many changes in real time. If our list had millions of titles, we'd better use a binary tree.

Such a tree might look something like this:

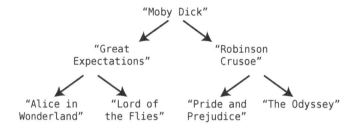

Now, we've already covered how to search, insert, and delete data from a binary tree. We mentioned, though, that we also want to be able to print the entire list of book titles in alphabetical order. How can we do that?

Firstly, we need the ability to visit every single node in the tree. The process of visiting every node in a data structure is known as *traversing* the data structure.

Secondly, we need to make sure that we traverse the tree in alphabetically ascending order so that we can print out the list in that order. There are multiple ways to traverse a tree, but for this application, we will perform what is known as *inorder traversal*, so that we can print out each title in alphabetical order.

Recursion is a great tool for performing inorder traversal. We'll create a recursive function called traverse that can be called on a particular node. The function then performs the following steps:

1. Call itself (traverse) on the node's left child if it has one.

2. Visit the node. (For our book title app, we print the value of the node at this step.)

3. Call itself (traverse) on the node's right child if it has one.

For this recursive algorithm, the base case is when a node has no children, in which case we simply print the node's title but do not call traverse again.

If we called traverse on the "Moby Dick" node, we'd visit all the nodes of the tree in the order shown on the diagram on page 165.

And that's exactly how we can print out our list of book titles in alphabetical order. Note that tree traversal is O(N), since by definition, traversal visits every node of the tree.

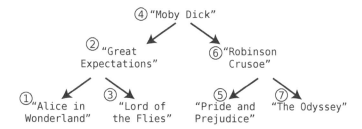

Here's a Python traverse_and_print function that works for our list of book titles:

```
def traverse_and_print(node):
    if node is None:
        return
    traverse_and_print(node.leftChild)
    print(node.value)
    traverse_and_print(node.rightChild)
```

Wrapping Up

Binary trees are a powerful node-based data structure that provides order maintenance, while also offering fast search, insertion, and deletion. They're more complex than their linked list cousins, but offer tremendous value.

It is worth mentioning that in addition to binary trees, there are many other types of tree-based data structures as well. Heaps, B-trees, red-black trees, and 2-3-4 trees, as well as many other trees each have their own use for specialized scenarios.

In the next chapter, we'll encounter yet another node-based data structure that serves as the heart of complex applications such as social networks and mapping software. It's powerful yet nimble, and it's called a graph.

CHAPTER 13

Connecting Everything with Graphs

Let's say that we're building a social network such as Facebook. In such an application, many people can be "friends" with one another. These friendships are mutual, so if Alice is friends with Bob, then Bob is also friends with Alice.

How can we best organize this data?

One very simple approach might be to use a two-dimensional array that stores the list of friendships:

```
relationships = [
  ["Alice", "Bob"],
  ["Bob", "Cynthia"],
  ["Alice", "Diana"],
  ["Bob", "Diana"],
  ["Elise", "Fred"],
  ["Diana", "Fred"],
  ["Fred", "Alice"]
]
```

Unfortunately, with this approach, there's no quick way to see who Alice's friends are. We'd have to inspect each relationship within the array, and check to see whether Alice is contained within the relationship. By the time we get through the entire array, we'd create a list of all of Alice's friends (who happen to be Bob, Diana, and Fred). We'd also perform the same process if we wanted to simply check whether Elise was Alice's friend.

Based on the way we've structured our data, searching for Alice's friends would have an efficiency of O(N), since we need to inspect every relationship in our database.

But we can do much, much better. With a data structure known as a *graph*, we can find each of Alice's friends in just O(1) time.

Graphs

A *graph* is a data structure that specializes in relationships, as it easily conveys how data is connected.

Here is a visualization of our Facebook network:

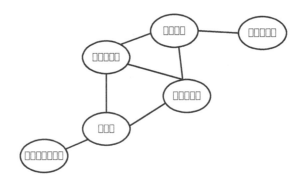

Each person is represented by a node, and each line indicates a friendship with another person. In graph jargon, each node is called a *vertex*, and each line is called an *edge*. Vertices that are connected by an edge are said to be *adjacent* to each other.

There are a number of ways that a graph can be implemented, but one of the simplest ways is using a hash table (see *Blazing Fast Lookup with Hash Tables*). Here's a bare-bones Ruby implementation of our social network:

```
friends = {
  "Alice" => ["Bob", "Diana", "Fred"],
  "Bob" => ["Alice", "Cynthia", "Diana"],
  "Cynthia" => ["Bob"],
  "Diana" => ["Alice", "Bob", "Fred"],
  "Elise" => ["Fred"],
  "Fred" => ["Alice", "Diana", "Elise"]
}
```

With a graph, we can look up Alice's friends in O(1), because we can look up the value of any key in a hash table in one step:

```
friends["Alice"]
```

With Twitter, in contrast to Facebook, relationships are *not* mutual. That is, Alice can follow Bob, but Bob doesn't necessarily follow Alice. Let's construct a new graph (shown at right) that demonstrates who follows whom.

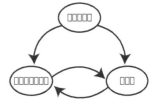

In this example, the arrows indicate the direction of the relationship. Alice follows both Bob and Cynthia, but no one follows Alice. Bob and Cynthia follow each other.

Using our hash table approach, we'd use the following code:

```
followees = {
  "Alice" => ["Bob", "Cynthia"],
  "Bob" => ["Cynthia"],
  "Cynthia" => ["Bob"]
}
```

While the Facebook and Twitter examples are similar, the nature of the relationships in each example are different. Because relationships in Twitter are one-directional, we use arrows in our visual implementation, and such a graph is known as a *directed graph*. In Facebook, where the relationships are mutual and we use simple lines, the graph is called a *non-directed graph*.

While a simple hash table can be used to represent a graph, a more robust object-oriented implementation can be used as well.

Here's a more robust implementation of a graph, using Ruby:

```
class Person
  attr_accessor :name, :friends
  def initialize(name)
    @name = name
    @friends = []
  end
  def add_friend(friend)
    @friends << friend
  end
end
```

With this Ruby class, we can create people and establish friendships:

```
mary = Person.new("Mary")
peter = Person.new("Peter")

mary.add_friend(peter)
peter.add_friend(mary)
```

Breadth-First Search

LinkedIn is another popular social network that specializes in business relationships. One of its well-known features is that you can determine your second- and third-degree connections in addition to your direct network.

In the diagram on the right, Alice is connected directly to Bob, and Bob is connect-
ed directly to Cynthia. However, Alice is not connected directly to Cynthia. Since Cynthia is connected to Alice by way of Bob, Cynthia is said to be Alice's second-degree connection.

If we wanted to find Alice's *entire* network, including her indirect connections, how would we go about that?

There are two classic ways to traverse a graph: *breadth-first search* and *depth-first search*. We'll explore breadth-first search here, and you can look up depth-first search on your own. Both are similar and work equally well for most cases, though.

The breadth-first search algorithm uses a queue (see *Crafting Elegant Code*), which keeps track of which vertices to process next. At the very beginning, the queue only contains the starting vertex (Alice, in our case). So, when our algorithm begins, our queue looks like this:

[Alice]

We then process the Alice vertex by removing it from the queue, marking it as having been "visited," and designating it as the current vertex. (This will all become clearer when we walk through the example shortly.)

We then follow three steps:

1. Visit each vertex adjacent to the current vertex. If it has not yet been visited, mark it as visited, and add it to a queue. (We do *not* yet make it the current vertex, though.)

2. If the current vertex has no unvisited vertices adjacent to it, remove the next vertex from the queue and make it the current vertex.

3. If there are no more unvisited vertices adjacent to the current vertex, and there are no more vertices in the queue, the algorithm is complete.

Let's see this in action. Here is Alice's LinkedIn network as shown in the top diagram on page 171. We start by making Alice the *current vertex*. To indicate in our diagram that she is the current vertex, we surround her with lines. We also put a check mark by her to indicate that she has been visited.

We then continue the algorithm by visiting an unvisited adjacent vertex—in this case, Bob. We add a check mark by his name too as shown in the second diagram on page 171.

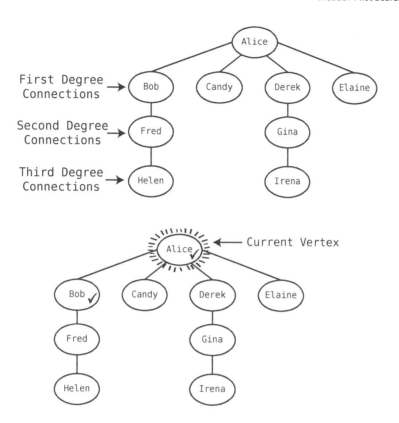

We also add Bob to our queue, so our queue is now: [Bob]. This indicates that we still have yet to make Bob the current vertex. Note that while Alice is the *current vertex*, we've still been able to *visit* Bob.

Next, we check whether Alice—the current vertex—has any other unvisited adjacent vertices. We find Candy, so we mark that as visited:

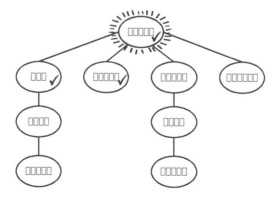

Our queue now contains [Bob, Candy].

Alice still has Derek as an unvisited adjacent vertex, so we visit it:

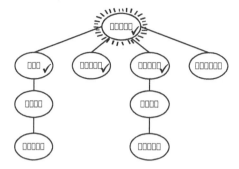

The queue is now [Bob, Candy, Derek].

Alice has one more unvisited adjacent connection, so we visit Elaine:

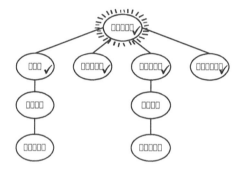

Our queue is now [Bob, Candy, Derek, Elaine].

Alice—our current vertex—has no more unvisited adjacent vertices, so we move on to the second rule of our algorithm, which says that we remove a vertex from the queue, and make it our current vertex. Recall from *Crafting Elegant Code* that we can only remove data from the *beginning* of a queue, so that would be Bob.

Our queue is now [Candy, Derek, Elaine], and Bob is our current vertex:

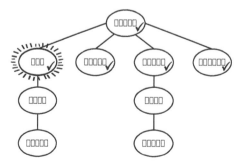

We now return to the first rule, which asks us to find any unvisited adjacent vertices of the current vertex. Bob has one such vertex—Fred—so we mark it as visited and add it to our queue:

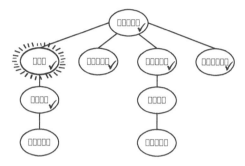

Our queue now contains [Candy, Derek, Elaine, Fred].

Since Bob has no more unvisited adjacent vertices, we remove the next vertex from the queue—Candy—and make it the current vertex:

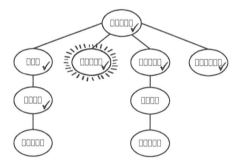

However, Candy has no unvisited adjacent vertices. So we grab the next item off the queue—Derek—leaving our queue as [Elaine, Fred]:

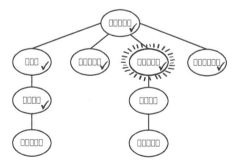

Derek has Gina as an unvisited adjacent connection, so we mark Gina as visited:

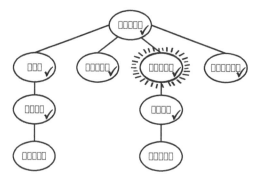

The queue is now [Elaine, Fred, Gina].

Derek has no more adjacent vertices to visit, so we take Elaine off the queue and make her the current vertex:

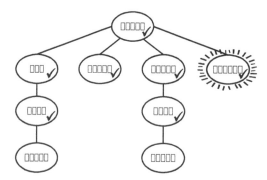

Elaine has no adjacent vertices that have not been visited, so we take Fred off the queue:

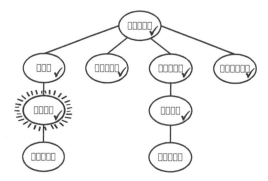

At this point, our queue is [Gina].

Now, Fred has one person to visit—Helen—so we mark her as visited and add her to the queue, making it [Gina, Helen]:

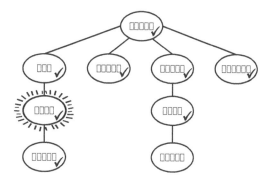

Since Fred has no more unvisited connections, we take Gina off the queue, making her the current vertex:

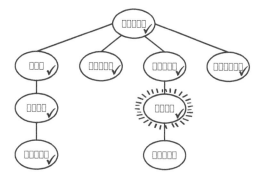

Our queue now only contains Helen: [Helen].

Gina has one vertex to visit—Irena:

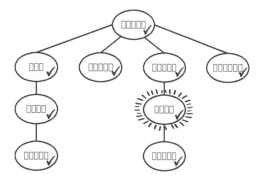

Our queue currently contains [Helen, Irena].

Gina has no more connections to visit, so we take Helen from the queue, making her the current vertex, and leaving our queue with [Irena]. Helen has nobody to visit, so we take Irena from the queue and make her the current vertex. Now, Irena also has no vertices to visit, so since the queue is empty—we're done!

Let's add a display_network method to our Person class that uses breadth-first search to display the names of a person's entire network:

```ruby
class Person
  attr_accessor :name, :friends, :visited

  def initialize(name)
    @name = name
    @friends = []
    @visited = false
  end

  def add_friend(friend)
    @friends << friend
  end

  def display_network
    # We keep track of every node we ever visit, so we can reset their
    # 'visited' attribute back to false after our algorithm is complete:
    to_reset = [self]

    # Create the queue. It starts out containing the root vertex:
    queue = [self]
    self.visited = true

    while queue.any?
      # The current vertex is whatever is removed from the queue
      current_vertex = queue.shift
      puts current_vertex.name

      # We add all adjacent vertices of the current vertex to the queue:
      current_vertex.friends.each do |friend|
        if !friend.visited
          to_reset << friend
          queue << friend
          friend.visited = true
        end
      end
    end

    # After the algorithm is complete, we reset each node's 'visited'
    # attribute to false:
    to_reset.each do |node|
      node.visited = false
    end
  end
end
```

To make this work, we've also added the visited attribute to the Person class that keeps track of whether the person was visited in the search.

The efficiency of breadth-first search in our graph can be calculated by breaking down the algorithm's steps into two types:

- We remove a vertex from the queue to designate it as the current vertex.
- For each current vertex, we visit each of its adjacent vertices.

Now, each vertex ends up being removed from the queue once. In Big O Notation, this is called O(V). That is, for V vertices in the graph, there are V removals from the queue.

Why don't we simply call this O(N), with N being the number of vertices? The answer is because in this (and many graph algorithms), we also have *additional* steps that process not just the vertices themselves, but also the *edges*, as we'll now explain.

Let's examine the number of times we visit each of the *adjacent vertices* of a current vertex.

Let's zero in on the step where Bob is the current vertex.

At this step, this code runs:

```
current_vertex.friends.each do |friend|
  if !friend.visited
    queue << friend
    friend.visited = true
  end
end
```

That is, we visit *every* vertex adjacent to Bob. This doesn't just include Fred, but it also includes Alice! While we don't queue Alice at this step because her vertex was already visited, she still takes up another step in our each loop.

If you walk through the steps of breadth-first search again carefully, you'll see that the number of times we visit adjacent vertices is twice the number of edges in the graph. This is because each edge connects two vertices, and for every vertex, we check all of its adjacent vertices. So each edge gets used twice.

So, for E edges, we check adjacent vertices 2E times. That is, for E edges in the graph, we check twice that number of adjacent vertices. However, since Big O ignores constants, we just write it as O(E).

Since there are O(V) removals from the queue, and O(E) visits, we say that breadth-first search has an efficiency of O(V + E).

Graph Databases

Because graphs are so versatile at handling relationships, there are actually databases that store data in the form of a graph. Traditional relational databases (databases that store data as columns and rows) can also store such data, but let's compare their performance against graph databases for dealing with something like a social network.

Let's say that we have a social network in which there are five friends who are all connected to each other. These friends are Alice, Bob, Cindy, Dennis, and Ethel. A graph database that will store their personal information may look something like this:

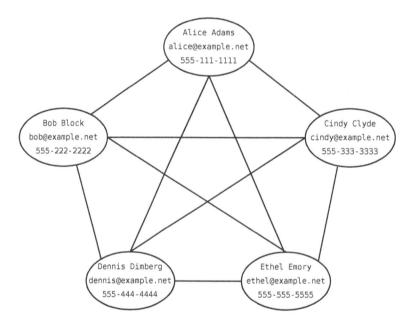

We can use a relational database to store this information as well. This would likely use two tables—one for storing the personal info of each user, and one for storing the relationships between the friends. Here is the Users table:

Users Table

id	firstname	lastname	email	phone
1	Alice	Adams	alice@example.net	555-111-1111
2	Bob	Block	bob@example.net	555-222-2222
3	Cindy	Clyde	cindy@example.net	555-333-3333
4	Dennis	Dimberg	dennis@example.net	555-444-4444
5	Ethel	Emory	ethel@example.net	555-555-5555

We would use a separate Friendships table to keep track of who is friends with whom:

Friendships Table

user_id	friend_id
1	2
1	3
1	4
1	5
2	1
2	3
2	4
2	5
3	1
3	2
3	4
3	5
4	1
4	2
4	3
4	5
5	1
5	2
5	3
5	4

← Alice is friends with Bob.

We'll avoid getting too deep into database theory, but note how this Friendships table uses the ids of each user to represent each user.

Assume that our social network has a feature that allows each user to see all the personal information about their friends. If Cindy were requesting this info, that would mean she'd like to see everything about Alice, Bob, Dennis, and Ethel, including their email addresses and phone numbers.

Let's see how Cindy's request would be executed if our application was backed by a relational database. First, we would have to look up Cindy's id in the Users table:

Users Table

id	firstname	lastname	email	phone
3	Cindy	Clyde	cindy@example.net	555-333-3333

↑ Cindy's id

Then, we'd look for all the rows in the Friendships table where the user id is 3:

Friendships Table

user_id	friend_id
3	1
3	2
3	4
3	5

↑ Cindy   ↑ Cindy's Friends

We now have a list of the ids of all of Cindy's friends: [1, 2, 4, 5].

With this id list, we need to return to the Users table to find each row with the appropriate id. The speed at which the computer can find each row in the Users table will be approximately O(log N). This is because the database maintains the rows in order of their ids, and the database can then use binary search to find each row. (This explanation applies to certain relational databases; other processes may apply to other relational databases.)

Since Cindy has four friends, the computer needs to perform O(log N) four times to pull all the friends' personal data. To say this more generally, for M friends, the efficiency of pulling their information is O(M log N). That is, for each friend, we run a search that takes log N steps.

Contrast this with getting the scoop on Cindy's friends if our application was backed by a graph database. In this case, once we've located Cindy in our database, it takes just *one step* to find one friend's info. This is because each vertex in the database contains all the information of its user, so we simply traverse the edges from Cindy to each of her friends. This takes a grand total of four steps as shown on page 181.

With a graph database, for N friends, pulling their data takes O(N) steps. This is a significant improvement over the efficiency of O(M log N) that a relational database would provide.

Neo4j[1] is an example of a popular open source graph database. I encourage you to visit their website for further resources about graph databases. Other examples of open source graph databases include ArangoDB[2] and Apache Giraph[3].

---

1. http://neo4j.com
2. https://www.arangodb.com/
3. http://giraph.apache.org/

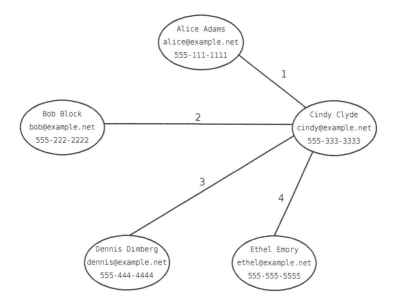

Note that graph databases aren't *always* the best solution for a given application. You'll need to carefully assess each application and its needs.

Weighted Graphs

Another type of graph is one known as a *weighted graph*. A weighted graph is like a regular graph but contains additional information about the edges in the graph.

Here's a weighted graph that represents a basic map of a few major cities in the United States:

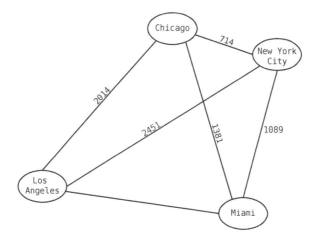

In this graph, each edge is accompanied by a number that represents the distance in miles between the cities that the edge connects. For example, there are 714 miles between Chicago and New York City.

It's also possible to have *directional* weighted graphs. In the following example, we can see that although a flight from Dallas to Toronto is $138, a flight from Toronto to Dallas is $216:

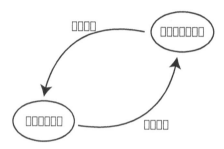

To add weights to our graph, we need to make a slight modification to our Ruby implementation. Specifically, we'll be using a hash table rather than an array to represent the adjacent nodes. In this case, each vertex will be represented by a City class:

```ruby
class City

  attr_accessor :name, :routes

  def initialize(name)
    @name = name
    # For the adjacent vertices, we are now using a hash table
    # instead of an array:
    @routes = {}
  end

  def add_route(city, price)
    @routes[city] = price
  end

end
```

Now, we can create cities and routes with prices:

```ruby
dallas = City.new("Dallas")
toronto = City.new("Toronto")

dallas.add_route(toronto, 138)
toronto.add_route(dallas, 216)
```

Let's use the power of the weighted graph to solve something known as the *shortest path problem*.

Here's a graph that demonstrates the costs of available flights between five different cities. (Don't ask me how airlines determine the prices of these things!)

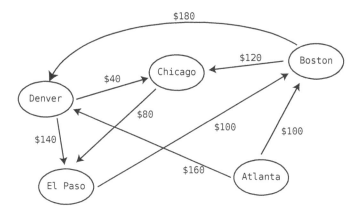

Now, say that I'm in Atlanta and want to fly to El Paso. Unfortunately, there's no direct route at this time. However, I can get there if I'm willing to fly through other cities. For example, I can fly from Atlanta to Denver, and from Denver to El Paso. That route would set me back $300. However, if you look closely, you'll notice that there's a cheaper route if I go from Atlanta to Denver to Chicago to El Paso. Although I have to take an extra flight, I only end up shelling out $280.

In this context, the shortest path problem is this: how can I get from Atlanta to El Paso with the least amount of money?

Dijkstra's Algorithm

There are numerous algorithms for solving the shortest path problem, and one really interesting one was discovered by Edsger Dijkstra (pronounced "dike' struh") in 1959. Unsurprisingly, this algorithm is known as *Dijkstra's algorithm*.

Here are the rules of Dijkstra's algorithm (don't worry—they'll become clearer when we walk through our example):

1. We make the starting vertex our current vertex.

2. We check all the vertices adjacent to the current vertex and calculate and record the weights from the starting vertex to all known locations.

3. To determine the next current vertex, we find the *cheapest unvisited* known vertex that can be reached from our starting vertex.

4. Repeat the first three steps until we have visited every vertex in the graph.

Let's walk through this algorithm step by step.

To record the cheapest price of the routes from Atlanta to other cities, we will use a table as follows:

From Atlanta to:	Boston	Chicago	Denver	El Paso
	?	?	?	?

First, we make the starting vertex (Atlanta) the current vertex. All our algorithm has access to at this point is the current vertex and its connections to its adjacent vertices. To indicate that it's the current vertex, we'll surround it with lines. And to indicate that this vertex was once the current vertex, we'll put a check mark through it:

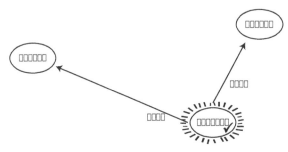

We next check all adjacent vertices and record the weights from the starting vertex (Atlanta) to all known locations. Since we can see that we can get from Atlanta to Boston for $100, and from Atlanta to Denver for $160, we'll record that in our table:

From Atlanta to:	Boston	Chicago	Denver	El Paso
	$100	?	$160	?

Next, we find the cheapest vertex that can be reached from Atlanta that has not yet been visited. We only know how to get to Boston and Denver from Atlanta at this point, and it's cheaper to get to Boston ($100) than it is to Denver ($160). So we make Boston our current vertex:

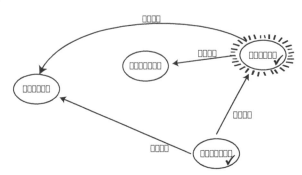

We now check both of the routes from Boston, and record all new data about the cost of the routes from *Atlanta—the starting vertex*—to all known locations. We can see that Boston to Chicago is $120. Now, we can also conclude that since Atlanta to Boston is $100, and Boston to Chicago is $120, the cheapest (and only) known route from Atlanta to Chicago is $220. We'll record this in our table:

From Atlanta to:	Boston	Chicago	Denver	El Paso
	$100	$220	$160	?

We also look at Boston's other route—which is to Denver—and that turns out to be $180. Now we have a new route from Atlanta to Denver: Atlanta to Boston to Denver. However, since this costs $280 while the direct flight from Atlanta to Denver is $160, we do not update the table, since we only keep the *cheapest known routes* in the table.

Now that we've explored all the outgoing routes from the current vertex (Boston), we next look for the unvisited vertex that is the cheapest to reach from Atlanta, which was our starting point. According to our table, Boston is still the cheapest known city to reach from Atlanta, but we've already checked it off. So that would make Denver the cheapest unvisited city, since it's only $160 from Atlanta, compared with Chicago, which is $220 from Atlanta. So Denver becomes our current vertex:

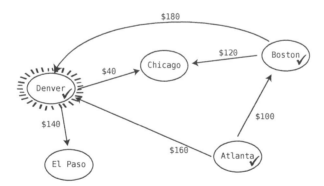

We now inspect the routes that leave Denver. One route is a $40 flight from Denver to Chicago. We can update our table since we now have a cheaper path from Atlanta to Chicago than before. Our table currently says that the cheapest route from Atlanta to Chicago is $220, but if we take Atlanta to Chicago by way of *Denver*, it will cost us just $200. We update our table accordingly:

From Atlanta to:	Boston	Chicago	Denver	El Paso
	$100	$200	$160	?

There's another flight out of Denver as well, and that is to the newly revealed city of El Paso. We now have to figure out the cheapest known path from Atlanta to El Paso, and that would be the $300 path from Atlanta to Denver to El Paso. We can now add this to our table as well:

From Atlanta to:	Boston	Chicago	Denver	El Paso
	$100	$200	$160	*$300*

We now have two unvisited vertices: Chicago and El Paso. Since the cheapest known path to Chicago from Atlanta ($200) is cheaper than the cheapest known path to El Paso from Atlanta ($300), we will next make Chicago the current vertex:

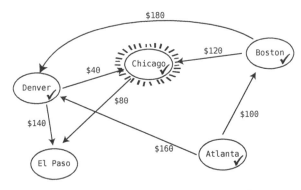

Chicago has just one outbound flight, and that is an $80 trip to El Paso. We now have a cheaper path from Atlanta to El Paso: Atlanta to Denver to Chicago to El Paso, which costs a total of $280. We update our table with this newfound data:

From Atlanta to:	Boston	Chicago	Denver	El Paso
	$100	$200	$160	*$280*

There's one known city left to make the current vertex, and that is El Paso:

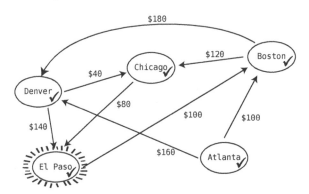

Dijkstra's Algorithm • 187

El Paso has only one outbound route, and that is a $100 flight to Boston. This data doesn't reveal any cheaper paths from Atlanta to anywhere, so we don't need to modify our table.

Since we have visited every vertex and checked it off, we now know every path from Atlanta to every other city. The algorithm is now complete, and our resulting table reveals the cheapest price of Atlanta to every other city on the map:

From Atlanta to:	Boston	Chicago	Denver	El Paso
	$100	$200	$160	$280

Here is a Ruby implementation of Dijkstra's algorithm:

We'll begin by creating a Ruby class representing a city. Each city is a node in a graph, which keeps track of its own name, and routes to adjacent cities:

```ruby
class City

  attr_accessor :name, :routes

  def initialize(name)
    @name = name
    # For the adjacent nodes, we are now using a hash table
    # instead of an array:
    @routes = {}
    # As an example, if this were Atlanta, its routes would be:
    # {boston => 100, denver => 160}
  end

  def add_route(city, price_info)
    @routes[city] = price_info
  end

end
```

We'll use the add_route method to set up the cities from our example:

```ruby
atlanta = City.new("Atlanta")
boston = City.new("Boston")
chicago = City.new("Chicago")
denver = City.new("Denver")
el_paso = City.new("El Paso")

atlanta.add_route(boston, 100)
atlanta.add_route(denver, 160)
boston.add_route(chicago, 120)
boston.add_route(denver, 180)
chicago.add_route(el_paso, 80)
denver.add_route(chicago, 40)
denver.add_route(el_paso, 140)
```

The code for Dijkstra's Algorithm is somewhat involved, so I've sprinkled comments liberally:

```ruby
def dijkstra(starting_city, other_cities)
  # The routes_from_city hash table below holds the data of all price_infos
  # from the given city to all other destinations, and the city which it
  # took to get there:
  routes_from_city = {}
  # The format of this data is:
  # {city => [price, other city which immediately precedes this city
  # along the path from the original city]}

  # In our example this will end up being:
  # {atlanta => [0, nil], boston => [100, atlanta], chicago => [200, denver],
  # denver => [160, atlanta], el_paso => [280, chicago]}

  # Since it costs nothing to get to the starting city from the starting city:
  routes_from_city[starting_city] = [0, starting_city]

  # When initializing our data, we set up all other cities having an
  # infinite cost - since the cost and the path to get to each other city
  # is currently unknown:
  other_cities.each do |city|
    routes_from_city[city] = [Float::INFINITY, nil]
  end
  # In our example, our data starts out as:
  # {atlanta => [0, nil], boston => [Float::INFINITY, nil],
  # chicago => [Float::INFINITY, nil],
  # denver => [Float::INFINITY, nil], el_paso => [Float::INFINITY, nil]}

  # We keep track of which cities we visited in this array:
  visited_cities = []

  # We begin visiting the starting city by making it the current_city:
  current_city = starting_city

  # We launch the heart of the algorithm, which is a loop that visits
  # each city:
  while current_city

    # We officially visit the current city:
    visited_cities << current_city

    # We check each route from the current city:
    current_city.routes.each do |city, price_info|
      # If the route from the starting city to the other city
      # is cheaper than currently recorded in routes_from_city, we update it:
      if routes_from_city[city][0] > price_info +
          routes_from_city[current_city][0]
        routes_from_city[city] =
          [price_info + routes_from_city[current_city][0], current_city]
      end
    end
```

```
      # We determine which city to visit next:
      current_city = nil
      cheapest_route_from_current_city = Float::INFINITY
      # We check all available routes:
      routes_from_city.each do |city, price_info|
      # if this route is the cheapest from this city, and it has not yet been
      # visited, it should be marked as the city we'll visit next:
        if price_info[0] < cheapest_route_from_current_city &&
            !visited_cities.include?(city)
          cheapest_route_from_current_city = price_info[0]
          current_city = city
        end
      end
    end

    return routes_from_city
end
```

We can run this method as follows:

```
routes = dijkstra(atlanta, [boston, chicago, denver, el_paso])
routes.each do |city, price_info|
  p "#{city.name}: #{price_info[0]}"
end
```

Although our example here focused on finding the cheapest flight, the same exact approach can be used for mapping and GPS technology. If the weights on each edge represented how fast it would take to drive from each city to the other rather than the price, we'd just as easily use Dijkstra's algorithm to determine which route you should take to drive from one place to another.

Wrapping Up

We're almost at the end of our journey, as this chapter represents the last significant data structure that we'll encounter in this book. We've seen that graphs are extremely powerful tools for dealing with data involving relationships, and in addition to making our code fast, they can also help solve tricky problems.

Along our travels, our primary focus has been on how fast our code will run. That is, we've been measuring how efficient our code performs in terms of time, and we've been measuring that in terms of counting the number of steps that our algorithms take.

However, efficiency can be measured in other ways. In certain situations, there are greater concerns than speed, and we might care more about how much *memory* a data structure or algorithm might consume. In the next chapter, we'll learn how to analyze the efficiency of our code in terms of *space*.

CHAPTER 14

Dealing with Space Constraints

Throughout this book, when analyzing the efficiency of various algorithms, we've focused exclusively on their *time complexity*—that is, how fast they run. There are situations, however, where we need to measure algorithm efficiency by another measure known as *space complexity*, which is how much memory an algorithm consumes.

Space complexity becomes an important factor when memory is limited. This can happen when programming for small hardware devices that only contain a relatively small amount of memory or when dealing with very large amounts of data, which can quickly fill even large memory containers.

In a perfect world, we'd always use algorithms that are both quick *and* consume a small amount of memory. However, there are times where we're faced with choosing between the speedy algorithm or the memory-efficient algorithm—and we need to look carefully at the situation to make the right choice.

Big O Notation as Applied to Space Complexity

Interestingly, computer scientists use Big O Notation to describe space complexity just as they do to describe time complexity.

Until now, we've used Big O Notation to describe an algorithm's speed as follows: for N elements of data, an algorithm takes a relative number of operational steps. For example, $O(N)$ means that for N data elements, the algorithm takes N steps. And $O(N^2)$ means that for N data elements, the algorithm takes N^2 steps.

Big O can similarly be used to describe how much space an algorithm takes up: *for N elements of data, an algorithm consumes a relative number of additional data elements in memory.* Let's take a simple example.

Let's say that we're writing a JavaScript function that accepts an array of strings, and returns an array of those strings in ALL CAPS. For example, the function would accept an array like ["amy", "bob", "cindy", "derek"] and return ["AMY", "BOB", "CINDY", "DEREK"]. Here's one way we can write this function:

```javascript
function makeUpperCase(array) {
  var newArray = [];
  for(var i = 0; i < array.length; i++) {
    newArray[i] = array[i].toUpperCase();
  }
  return newArray;
}
```

In this makeUpperCase() function, we accept an array called array. We then create a *brand-new array* called newArray, and fill it with uppercase versions of each string from the original array.

By the time this function is complete, we have two arrays floating around in our computer's memory. We have array, which contains ["amy", "bob", "cindy", "derek"], and we have newArray, which contains ["AMY", "BOB", "CINDY", "DEREK"].

When we analyze this function in terms of space complexity, we can see that it accepts an array with N elements, and creates a brand-new array that also contains N elements. Because of this, we'd say that the makeUpperCase() function has a space efficiency of O(N).

The way this appears on the following graph should look quite familiar:

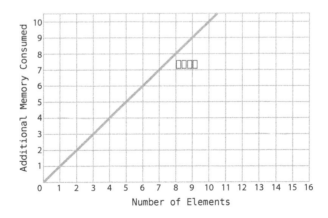

Note that this graph is identical to the way we've depicted O(N) in the graphs from previous chapters, with the exception that the vertical axis now represents memory rather than speed.

Now, let's present an alternative makeUpperCase() function that is more memory efficient:

```
function makeUpperCase(array) {
  for(var i = 0; i < array.length; i++) {
    array[i] = array[i].toUpperCase();
  }
  return array;
}
```

In this second version, we do not create any new variables or new arrays. In fact, we have not consumed any new memory at all. Instead, we modify each string within the original array in place, making them uppercase one at a time. We then return the modified array.

Since this function does not consume any memory in addition to the original array, we'd describe the space complexity of this function as being O(1). Remember that by time complexity, O(1) represents that the speed of an algorithm is constant no matter how large the data. Similarly, by space complexity, O(1) means that the memory consumed by an algorithm is constant no matter how large the data.

In our case, the algorithm consumes the same amount of additional space (zero!) no matter whether the original array contains four elements or one hundred. Because of this, our new version of makeUpperCase() is said to have a space efficiency of O(1).

It's important to reiterate that in this book, we judge the space complexity based on *additional* memory consumed—known as *auxiliary space*—meaning that we don't count the original data. Even in this second version, we do have an input of array, which contains N elements of memory. However, since this function does not consume any memory *in addition* to the original array, it is O(1).

(There are some references that include the original input when calculating the space complexity, and that's fine. We're not including it, and whenever you see space complexity described, you need to determine whether it's including the original input.)

Let's now compare the two versions of makeUpperCase() in both time and space complexity:

Version	Time complexity	Space complexity
Version #1	O(N)	O(N)
Version #2	O(N)	O(1)

Both versions are O(N) in time complexity, since they take N steps for N data elements. However, the second version is more memory efficient, being O(1) in space complexity compared to the first version's O(N).

This is a pretty strong case that version #2 is preferable to version #1.

Trade-Offs Between Time and Space

In *Speeding Up Your Code with Big O*, we wrote a JavaScript function that checked whether an array contained any duplicate values. Our first version looked like this:

```
function hasDuplicateValue(array) {
    for(var i = 0; i < array.length; i++) {
        for(var j = 0; j < array.length; j++) {
            if(i !== j && array[i] == array[j]) {
                return true;
            }
        }
    }
    return false;
}
```

It uses nested for loops, and we pointed out that it has a time complexity of $O(N^2)$.

We then created a more efficient version, which looked like this:

```
function hasDuplicateValue(array) {
    var existingNumbers = [];
    for(var i = 0; i < array.length; i++) {
        if(existingNumbers[array[i]] === undefined) {
            existingNumbers[array[i]] = 1;
        } else {
            return true;
        }
    }
    return false;
}
```

This second version creates an array called existingNumbers, and for each number we encounter in array, we find the corresponding index in existingNumbers and fill it with a 1. If we encounter a number and already find a 1 in the corresponding index within existingNumbers, we know that this number already exists and that we've therefore found a duplicate value.

We declared victory with this second version, pointing to its time complexity of O(N) compared with the first version's $O(N^2)$. Indeed, from the perspective of time alone, the second version is much faster.

However, when we take *space* into account, we find that this second version has a disadvantage compared with the first version. The first version does not consume any additional memory beyond the original array, and therefore has a space complexity of O(1). On the other hand, this second version *creates a brand-new array* that is the same size as the original array, and therefore has a space complexity of O(N).

Let's look at the complete contrast between the two versions of hasDuplicateValue():

Version	Time complexity	Space complexity
Version #1	O(N^2)	O(1)
Version #2	O(N)	O(N)

We can see that version #1 takes up less memory but is slower, while version #2 is faster but takes up more memory. How do we decide which to choose?

The answer, of course, is that it depends on the situation. If we need our application to be blazing fast, and we have enough memory to handle it, then version #2 might be preferable. If, on the other hand, speed isn't critical, but we're dealing with a hardware/data combination where we need to consume memory sparingly, then version #1 might be the right choice. Like all technology decisions, when there are trade-offs, we need to look at the big picture.

Parting Thoughts

You've learned a lot in this journey. Most importantly, you've learned that the analysis of data structures and algorithms can dramatically affect your code—in speed, memory, and even elegance.

What you can take away from this book is a framework for making educated technology decisions. Now, computing contains many details, and while something like Big O Notation may suggest that one approach is better than another, other factors can also come in to play. The way memory is organized within your hardware and the way that your computer language of choice implements things under the hood can also affect how efficient your code may be. Sometimes what you *think* is the most efficient choice may not be due to various external reasons.

Because of this, it is always best to *test* your optimizations with benchmarking tools. There are many excellent software applications out there that can measure the speed and memory consumption of your code. The knowledge in this book will point you in the right direction, and the benchmarking tools will confirm whether you've made the right choices.

I hope that you also take away from this book that topics like these—which seem so complex and esoteric—are just a combination of simpler, easier concepts that are within your grasp. Don't be intimidated by resources that make a concept seem difficult simply because they don't explain it well—you can always find a resource that explains it better.

The topic of data structures and algorithms is broad and deep, and we've only just scratched the surface. There is so much more to learn—but with the foundation that you now have, you'll be able to do so. Good luck!

Index

A

algorithms, 17, *see also* examples; operations
   binary search, 21–26
   Bubble Sort, 37–45
   Dijkstra's algorithm, 183–189
   Insertion Sort, 63–74
   linear search, 7–8, 13, 20–21, 24–26
   Quickselect, 128–131
   Quicksort, 113–128
   Selection Sort, 51–59
array-based sets, 12, *see also* sets
arrays, 2–12, *see also* ordered arrays
   definition, 2–4
   deleting items from, 3, 11–12
   duplicates, checking for, 45–49
   index of, 2
   inserting items in, 3, 9–11
   intersection of two arrays, 74–76
   reading, 3–6
   searching, 3, 7–8
associative arrays, *see* hash tables
average-case scenario, 71–74

B

base case, for recursion, 105
Big O Notation, 27–36, *see also* time complexity (speed)
   constant time, 29–31
   constants ignored rule, 58–59
   highest order of N rule, 71
   linear time, 29–31, 47–49
   log time, 32–35
   quadratic time, 45–47
   space complexity using, 191–194
   when to use, 59–62
   worst-case scenario used by, 32
binary search, 21–26
   compared to linear search, 24–26
   definition, 21–24
   time complexity of, 32
binary trees, 149–165
   book titles example using, 163–165
   compared to ordered arrays, 154, 156
   definition, 149–152
   deleting items from, 157–163
   inserting items in, 154–157
   levels of, 150
   parents and children in, 150
   root of, 150
   searching, 152–154
   traversing, 164–165
   well-balanced, 156
book titles example, 163–165
books, *Introduction to Algorithms* (Cormen, Leiserson, Rivest, and Stein), 29
breadth-first searches, 169–177
Bubble Sort, 37–45
   compared to Insertion Sort, 71
   compared to Selection Sort, 58–59, 71
   definition, 37–43
   time complexity of, 43–45

C

call stack, for recursion, 108–109
children, in binary tree, 150
collisions, with hash tables, 82–86
constant time, 29–31
constants, ignored in Big O Notation, 58–59

D

data, definition, 1
data structures
   arrays, 2–12
   binary trees, 149–165
   definition, 1
   graphs, 168–189
   hash tables, 77–89
   linked lists, 133–147
   operations on, 3
   ordered arrays, 18–26
   performance affected by, 1–2
   queues, 98–101
   sets, 12–15
   stacks, 92–98
data, temporary, *see* queues; stacks

databases, for graphs, 178–181
delete operation
    arrays, 3, 11–12
    binary trees, 157–163
    linked lists, 140–142
    sets, 13
depth-first searches, 170
dictionaries, see hash tables
Dijkstra's algorithm, 183–189
directed graphs, 169
directional weighted graphs, 182
doubly linked lists, 143–147
duplicates, checking for, 45–49, 87–88

E

edges, in graphs, 168
efficiency, see space complexity; time complexity
electronic voting machine example, 88–89
examples
    book titles, 163–165
    duplicates, checking for, 45–49, 87–88
    electronic voting machine, 88–89
    factorials, calculating, 105–109
    filesystem traversal, 110–111
    JavaScript linter, 93–98
    printer manager, 100–101
    shortest path problem, 182–189
    social networks, 167–181
    thesaurus using hash table, 79–81

F

Facebook example, 167–169
factorials, calculating, 105–109
FIFO (First In, First Out), 98
filesystem traversal example, 110–111

G

graphs, 168–189
    breadth-first searches of, 169–177
    databases used for, 178–181
    definition, 168–169
    depth-first searches of, 170
    directed, 169
    directional weighted, 182
    edges in, 168
    hash tables used for, 168
    non-directed, 169
    queues used in searching, 170
    shortest path problem using, 182–189
    social network examples using, 167–181
    traversing, 170
    vertices in, 168
    weighted, 181

H

hash functions, 78–79, 86
hash tables, 77–89
    collisions, handling, 82–86
    definition, 77–79
    duplicates in, checking for, 87–88
    electronic voting machine example using, 88–89
    graphs implemented using, 168
    keys for, 78
    load factor for, 86
    sets implemented using, 87–89
    size of, optimum, 85–86
    thesaurus example using, 79–81
    time complexity of, 84–86
    values for, 78
    weighted graphs using, 182
hashing, 78

I

index, of arrays, 2
insert operation
    arrays, 3, 9–11
    binary trees, 154–157
    linked lists, 138–140
    ordered arrays, 18–20
    sets, 13–15
Insertion Sort, 63–74
    compared Selection Sort, 71–74
    definition, 63–69
    time complexity of, 69–74
Introduction to Algorithms (Cormen, Leiserson, Rivest, and Stein), 29

J

JavaScript linter example, 93–98

L

levels, of binary tree, 150
LIFO (Last In, First Out), 93
linear search, 7–8, 20–21
    of arrays, 7–8
    compared to binary search, 24–26
    of ordered arrays, 20–21
    of sets, 13
    time complexity of, 28, 31
linear time, 29–31, 47–49
linked lists, 133–147
    compared to ordered arrays, 142
    definition, 133–135
    deleting items from, 140–142
    doubly linked lists, 143–147
    implementing, 135–136
    inserting items in, 138–140
    link in each node, 134
    reading items from, 136–137
    searching, 137–138
    when to use, 142
LinkedIn example, 169–177
linter example, for JavaScript, 93–98
load factor, for hash tables, 86
log time, 32–35
logarithms, 33–34
loops, recursion as alternative to, 103–104

M

maps, see hash tables
mathematical notation
    Big O Notation, 27–36
    used in this book, ix, 29
memory used, see space complexity

N

node-based data structures, *see* binary trees; graphs; linked lists
non-directed graphs, 169

O

"O" notation, *see* Big O Notation
online resources
    Big O Notation, 29
    open source graph databases, 180
    for this book, xii
    tree-based data structures, 165
operations, 3, *see also* algorithms; delete operation; insert operation; read operation; search operation; traversal operation
"Order of" notation, *see* Big O Notation
ordered arrays, 18–26, *see also* arrays
    compared to binary trees, 154, 156
    compared to linked lists, 142
    definition, 18–20
    inserting items in, 18–20
    searching, 20–26

P

parent, in binary tree, 150
partitioning, 113–117
    time complexity of, 123–124
performance, *see* time complexity
pivot, in partitioning, 113
printer manager example, 100–101

Q

quadratic time, 45–47
queues, 98–101
    adding items to, 99
    definition, 98–99
    doubly linked lists used for, 143–147
    FIFO operations for, 98
    graph searches using, 170
    printer manager example using, 100–101
    removing items from, 99
Quickselect, 128–131
Quicksort, 113–128
    definition, 118–123
    partitioning used by, 113–117
    time complexity of, 123–128
    worst-case scenario for, 126–128

R

read operation
    arrays, 3–6
    linked lists, 136–137
    sets, 13
recursion, 103–109
    as alternative to loops, 103–104
    base case for, 105
    call stack for, 108–109
    computer's implementation of, 108–109
    definition, 103
    factorials example using, 105–109
    filesystem traversal example using, 110–111
    Quickselect using, 128–131
    Quicksort using, 118–123
    reading code using, 105–108
    stack overflow from, 109
relational databases, compared to graph databases, 178–181
resources, *see* books; online resources
root, of binary tree, 150

S

search operation
    arrays, 3, 7–8
    binary search, 21–26
    binary trees, 152–154
    graphs, 169–177
    linear search, 7–8, 13, 20–21
    linked lists, 137–138
    ordered arrays, 20–26
    sets, 13
Selection Sort, 51–59
    compared to Insertion Sort, 71–74
    definition, 51–57
    time complexity of, 57–59
sets, 12–15
    definition, 12–13
    deleting items from, 13
    hash tables used as, 87–89
    inserting items in, 13–15
    reading, 13
    searching, 13
shortest path problem, 182–189
social network examples, 167–181
sorting algorithms
    Bubble Sort, 37–45
    Insertion Sort, 63–74
    Quicksort, 113–128
    Selection Sort, 51–59
space complexity, 191–195
    Big O Notation for, 191–194
    trade-offs with time complexity, 194–195
speed, *see* time complexity
stack overflow, from recursion, 109
stacks, 92–98
    call stack, for recursion, 108–109
    definition, 92–93
    JavaScript linter example using, 93–98
    LIFO operations for, 93
    popping items from, 93
    pushing items onto, 92
standard arrays, *see* arrays
steps, as efficiency measurement, 3, 28

T

temporary data, handling, *see* queues; stacks
thesaurus example, 79–81
time complexity (speed)
    average-case scenario for, 71–74
    Big O Notation for, 27–36
    binary compared to linear searches, 24–26
    binary search, 32
    binary tree deletions, 163
    binary tree insertions, 156
    binary tree searches, 154, 157

binary tree traversal, 164
Bubble Sort, 43–45
data structures affecting, 1–2
graph database searches, 180
graph searches, 177
hash tables, 84–86
Insertion Sort, 69–74
linear search, 28, 31
measuring in steps, 3, 28
ordered compared to standard arrays, 26

Quickselect, 130
Quicksort, 123–128
relational database searches, 180
Selection Sort, 57–59
trade-offs with space complexity, 194–195
worst-case scenario for, 32
traversal operation
 binary trees, 164–165
 filesystem, 110–111
 graphs, 170

tree-based data structures, 165, *see also* binary trees
Twitter example, 168–169

V

vertices, in graphs, 168

W

website resources, *see* online resources
weighted graphs, 181
worst-case scenario, 32

Thank you!

How did you enjoy this book? Please let us know. Take a moment and email us at support@pragprog.com with your feedback. Tell us your story and you could win free ebooks. Please use the subject line "Book Feedback."

Ready for your next great Pragmatic Bookshelf book? Come on over to https://pragprog.com and use the coupon code BUYANOTHER2018 to save 30% on your next ebook.

Void where prohibited, restricted, or otherwise unwelcome. Do not use ebooks near water. If rash persists, see a doctor. Doesn't apply to *The Pragmatic Programmer* ebook because it's older than the Pragmatic Bookshelf itself. Side effects may include increased knowledge and skill, increased marketability, and deep satisfaction. Increase dosage regularly.

And thank you for your continued support,

Andy Hunt, Publisher

Put the "Fun" in Functional

Elixir puts the "fun" back into functional programming, on top of the robust, battle-tested, industrial-strength environment of Erlang.

Programming Elixir 1.6

This book is *the* introduction to Elixir for experienced programmers, completely updated for Elixir 1.6 and beyond. Explore functional programming without the academic overtones (tell me about monads just one more time). Create concurrent applications, but get them right without all the locking and consistency headaches. Meet Elixir, a modern, functional, concurrent language built on the rock-solid Erlang VM. Elixir's pragmatic syntax and built-in support for metaprogramming will make you productive and keep you interested for the long haul. Maybe the time is right for the Next Big Thing. Maybe it's Elixir.

Dave Thomas
(410 pages) ISBN: 9781680502992. $47.95
https://pragprog.com/book/elixir16

Metaprogramming Elixir

Write code that writes code with Elixir macros. Macros make metaprogramming possible and define the language itself. In this book, you'll learn how to use macros to extend the language with fast, maintainable code and share functionality in ways you never thought possible. You'll discover how to extend Elixir with your own first-class features, optimize performance, and create domain-specific languages.

Chris McCord
(128 pages) ISBN: 9781680500417. $17
https://pragprog.com/book/cmelixir

More Functional Languages

Dig into domain modeling in F#, and learn all about Clojure on the Java VM.

Domain Modeling Made Functional

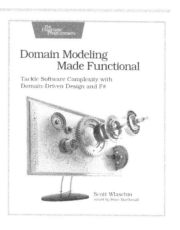

You want increased customer satisfaction, faster development cycles, and less wasted work. Domain-driven design (DDD) combined with functional programming is the innovative combo that will get you there. In this pragmatic, down-to-earth guide, you'll see how applying the core principles of functional programming can result in software designs that model real-world requirements both elegantly and concisely—often more so than an object-oriented approach. Practical examples in the open-source F# functional language, and examples from familiar business domains, show you how to apply these techniques to build software that is business-focused, flexible, and high quality.

Scott Wlaschin
(310 pages) ISBN: 9781680502541. $47.95
https://pragprog.com/book/swdddf

Programming Clojure, Third Edition

Drowning in unnecessary complexity, unmanaged state, and tangles of spaghetti code? In the best tradition of Lisp, Clojure gets out of your way so you can focus on expressing simple solutions to hard problems. Clojure cuts through complexity by providing a set of composable tools—immutable data, functions, macros, and the interactive REPL. Written by members of the Clojure core team, this book is the essential, definitive guide to Clojure. This new edition includes information on all the newest features of Clojure, such as transducers and specs.

Alex Miller with Stuart Halloway and Aaron Bedra
(302 pages) ISBN: 9781680502466. $49.95
https://pragprog.com/book/shcloj3

Seven in Seven

You need to learn at least one new language every year. Here are fourteen excellent suggestions to get started.

Seven Languages in Seven Weeks

You should learn a programming language every year, as recommended by *The Pragmatic Programmer*. But if one per year is good, how about *Seven Languages in Seven Weeks*? In this book you'll get a hands-on tour of Clojure, Haskell, Io, Prolog, Scala, Erlang, and Ruby. Whether or not your favorite language is on that list, you'll broaden your perspective of programming by examining these languages side-by-side. You'll learn something new from each, and best of all, you'll learn how to learn a language quickly.

Bruce A. Tate
(330 pages) ISBN: 9781934356593. $34.95
https://pragprog.com/book/btlang

Seven More Languages in Seven Weeks

Great programmers aren't born—they're made. The industry is moving from object-oriented languages to functional languages, and you need to commit to radical improvement. New programming languages arm you with the tools and idioms you need to refine your craft. While other language primers take you through basic installation and "Hello, World," we aim higher. Each language in *Seven More Languages in Seven Weeks* will take you on a step-by-step journey through the most important paradigms of our time. You'll learn seven exciting languages: Lua, Factor, Elixir, Elm, Julia, MiniKanren, and Idris.

Bruce Tate, Fred Daoud, Jack Moffitt, Ian Dees
(318 pages) ISBN: 9781941222157. $38
https://pragprog.com/book/7lang

Long Live the Command Line!

Use tmux and Vim for incredible mouse-free productivity.

tmux 2

Your mouse is slowing you down. The time you spend context switching between your editor and your consoles eats away at your productivity. Take control of your environment with tmux, a terminal multiplexer that you can tailor to your workflow. With this updated second edition for tmux 2.3, you'll customize, script, and leverage tmux's unique abilities to craft a productive terminal environment that lets you keep your fingers on your keyboard's home row.

Brian P. Hogan
(102 pages) ISBN: 9781680502213. $21.95
https://pragprog.com/book/bhtmux2

Practical Vim, Second Edition

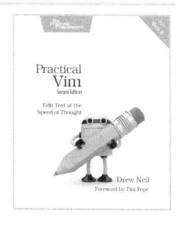

Vim is a fast and efficient text editor that will make you a faster and more efficient developer. It's available on almost every OS, and if you master the techniques in this book, you'll never need another text editor. In more than 120 Vim tips, you'll quickly learn the editor's core functionality and tackle your trickiest editing and writing tasks. This beloved bestseller has been revised and updated to Vim 8 and includes three brand-new tips and five fully revised tips.

Drew Neil
(354 pages) ISBN: 9781680501278. $29
https://pragprog.com/book/dnvim2

The Pragmatic Bookshelf

The Pragmatic Bookshelf features books written by developers for developers. The titles continue the well-known Pragmatic Programmer style and continue to garner awards and rave reviews. As development gets more and more difficult, the Pragmatic Programmers will be there with more titles and products to help you stay on top of your game.

Visit Us Online

This Book's Home Page
https://pragprog.com/book/jwdsal
Source code from this book, errata, and other resources. Come give us feedback, too!

Keep Up to Date
https://pragprog.com
Join our announcement mailing list (low volume) or follow us on twitter @pragprog for new titles, sales, coupons, hot tips, and more.

New and Noteworthy
https://pragprog.com/news
Check out the latest pragmatic developments, new titles and other offerings.

Save on the eBook

Save on the eBook versions of this title. Owning the paper version of this book entitles you to purchase the electronic versions at a terrific discount.

PDFs are great for carrying around on your laptop—they are hyperlinked, have color, and are fully searchable. Most titles are also available for the iPhone and iPod touch, Amazon Kindle, and other popular e-book readers.

Buy now at *https://pragprog.com/coupon*

Contact Us

Online Orders:	*https://pragprog.com/catalog*
Customer Service:	*support@pragprog.com*
International Rights:	*translations@pragprog.com*
Academic Use:	*academic@pragprog.com*
Write for Us:	*http://write-for-us.pragprog.com*
Or Call:	+1 800-699-7764

CPSIA information can be obtained
at www.ICGtesting.com
Printed in the USA
BVHW08s1013050718
520757BV00009B/31/P